Studies in Space Policy

Volume 19

Series Editor

European Space Policy Institute, Vienna, Austria

Edited by the European Space Policy Institute
Director: Jean-Jacques Tortora

The use of outer space is of growing strategic and technological relevance. The development of robotic exploration to distant planets and bodies across the solar system, as well as pioneering human space exploration in earth orbit and of the moon, paved the way for ambitious long-term space exploration. Today, space exploration goes far beyond a merely technological endeavour, as its further development will have a tremendous social, cultural and economic impact. Space activities are entering an era in which contributions of the humanities—history, philosophy, anthropology—, the arts, and the social sciences—political science, economics, law—will become crucial for the future of space exploration. Space policy thus will gain in visibility and relevance. The series Studies in Space Policy shall become the European reference compilation edited by the leading institute in the field, the European Space Policy Institute. It will contain both monographs and collections dealing with their subjects in a transdisciplinary way.

More information about this series at http://www.springer.com/series/8167

Annette Froehlich

Editor

Legal Aspects Around Satellite Constellations

Editor
Annette Froehlich (iD)
European Space Policy Institute
Vienna, Austria

ISSN 1868-5307   ISSN 1868-5315   (electronic)
Studies in Space Policy
ISBN 978-3-030-06027-5   ISBN 978-3-030-06028-2   (eBook)
https://doi.org/10.1007/978-3-030-06028-2

Library of Congress Control Number: 2019933358

This Springer imprint is published by the registered company Springer Nature Switzerland AG
The registered company address is: Gewerbestrasse 11, 6330 Cham, Switzerland

Preface

In view of the proliferating activities in the field of satellite constellations, the European Space Policy Institute (ESPI), the European Centre for Space Law (ECSL), and the German Aerospace Centre (DLR) invited students and young professionals worldwide to submit a paper on "Legal aspects around satellite constellations" as these are becoming increasingly important. Indeed, these groups of coordinated satellites evoke a wide range of interesting topics from various fields (environment, economy, security, licencing, control, etc.), which highlight the relevance of this topic.

The main focus was placed on the management of scarce orbital resources, including geostationary orbit (GSO) and low Earth orbit (LEO), in order to guarantee efficient use and equitable access while taking into account the needs and interests of the whole international community. Indeed, with the rise of private actors and their announced mega constellations, LEO seems to be particularly in high demand. Therefore, the frequency assignments should be management without discrimination to avoid a domination of LEO by some major space actors. In addition, even if private companies from established space-faring nations like the US currently intend to set up satellite constellations to provide Internet for the benefit of humanity, this should nevertheless not lead to a situation where relevant orbits are no longer available for new actors from emerging space countries.

With the emergence of private actors, coupled with the fact that States are no longer the primary actors, a review of the legal situation is furthermore envisaged to enhance legal certainty. Therefore, proposals include a harmonization of the legal regimes and the elaboration of flexible solutions, even a separate regime, through analysis of the link and interconnectivity between the legal telecommunication regimes provided by the International Telecommunication Union (ITU) and the use of outer space as guaranteed by the UN treaties in the context of satellite constellations. The flexible mechanism of the ITU may especially allow more timely action and an efficient system to avoid warehousing of frequency assignments. In addition, regulations to ensure protection from technical interference are of utmost importance, especially for mega constellations, to avoid interfering incidents.

Finally, while satellite constellations present new opportunities, they can also compound challenges related to space debris, especially with the emergence of small satellite constellations. Therefore, end-of-life disposal activities have to be envisaged for such an extensive amount of satellites, while existing guidelines in this regard are scrutinized, especially as to whether they are sufficient for further satellite constellation projects.

The sustained support of Mari Amanda Eldholm, Executive Secretary of the European Centre for Space Law (ECSL), was invaluable to the success of this project, and sincere gratitude and appreciation for her enthusiastic cooperation is hereby expressed.

Vienna, Austria   Annette Froehlich
October 2018

Contents

Chapter 1
An Equitable and Efficient Use of Outer Space and Its Resources and the Role of the UN, the ITU and States Parties

Margaux Morssink

Abstract The development of satellite constellations and large infrastructures has been coming the last few years. Although this advancement provides for a myriad of technological developments, private initiatives and opportunities, it also provides governments, institutions and private actors with regulatory and legal challenges.

The introduction of satellite constellations and infrastructures provides for different challenges of a various nature. This chapter focuses on the link between the legal regimes of telecommunications and the use of outer space in relation to satellite constellations. It explores how the scarcity of frequencies, which is a topic that is traditionally affiliated to (national and international) telecommunications law, will be a very relevant topic in relation to the advent of satellite constellations and space-law-related topics. It also raises the question how the role of administrations and supranational organisations can be filled in the light of these recent developments, and it attempts to address a couple of current issues at hand.

1.1 Introduction

Over the last couple of years, a lot of developments in regard to satellite constellations have taken place. Several companies announced that they were planning on launching satellite constellations to provide Internet to every place on Earth.[1] With these new developments, more and more satellites are and will be launched. This will increase the congestion of the lower Earth orbit and the geostationary orbit, as well as other orbits. Another factor in this development, apart from technological advancement, is the rise of privatisation and private actors in the space industry. As

---

The author has written this chapter in her personal capacity and its contents should not be attributed to her position at the Dutch Radio Communications Agency.

[1] See for example: SpaceNews, 'SpaceX, OneWeb detail constellation plans to Congress' (October 26, 2017) <https://spacenews.com/spacex-oneweb-detail-constellation-plans-to-congress/> accessed 29 September 2018.

M. Morssink (✉)
Agentschap Telecom (Dutch Radio Communications Agency), Groningen, The Netherlands

© Springer Nature Switzerland AG 2019

A. Froehlich (ed.), *Legal Aspects Around Satellite Constellations*, Studies in Space Policy 19, https://doi.org/10.1007/978-3-030-06028-2_1

a result, these private actors will have more control, which will make financial considerations more important. Apart from the perhaps obvious environmental considerations to take into account (congestion of orbits, space debris mitigation), issues related to satellite communications should not be overlooked. The increase of satellite constellations or large satellite infrastructures could possibly lead to congestion in specific frequency bands. In general, frequencies are perceived to be scarce resources. What happens to frequency filings if more and more satellite constellations are launched? How will this be regulated within the ITU? What will this mean for the bringing into use system of the ITU? As regulation of radio frequencies, especially in relation to mobile and broadcasting, is predominantly reserved to national governments, the question arises whether governments also have a role in the regulation of the use of frequencies in relation to satellite constellations. This chapter examines these questions and poses the statement that a uniform regime is required to avoid 'shopping' between States and to avoid spectrum warehousing.

This chapter examines the link between space and telecommunications law in the context of satellite constellations. On the surface, launching satellite constellations may seem like a predominantly space-law-related topic. For example, it raises questions on the environment (deorbiting, the sustainable use of outer space), security and economics. The rise of satellite constellations does, however, also raise questions that are predominantly related to the regime of telecommunications law. This chapter poses the statement that where the two domains of law used to be separate and have two separate regimes (the UN Space Treaties and the ITU Radio Regulations), they will become more and more interrelated. This is especially apparent with satellite constellations and large satellite infrastructures. In this area, the issue of congestion of outer space arises, in terms of both the space that the satellites take up and the frequency spectrum that is needed for these satellites.

The United Nations Office for Outer Space Affairs (UNOOSA) and the ITU published a document on the guidance of space object registration and frequency management for small and very small satellites. This is relevant in regard to the topic of satellite constellations, as some of the new initiatives concern small satellites launched into the low Earth orbit (LEO). In this document, certain requirements are explained. Reference is made to the fact that for the launch and operation of satellites, certain requirements exist, including 'notification and recording of the radio frequencies at the ITU, consideration of space debris mitigation measures in the design and operation of a satellite and registration of the satellite(s) with the Secretary-General of the United Nations'.[2] Thus, for the launch and operation of satellite constellations with small satellites,[3] clear regulation related to both regimes is needed.

---

[2] United Nations Office for Outer Space Affairs and the International Telecommunication Union UNCOPUOS (Legal Subcommittee) Fifty-fourth session 13-24 April 2015 'Guidance on Space Object Registration and Frequency Management for Small and Very Small Satellites' (13 April 2015) A/AC.105/C.2/2015/CRP.17.

[3] OneWeb is planning on launching a satellite constellation, consisting of more than 900 microsatellites.

1.2   Legal Regimes

1.2.1   Telecommunications

The International Telecommunication Union

The International Telecommunication Union (hereinafter 'ITU') was established in 1865 (it was at that time called the International Telegraph Union) and celebrated its 105th birthday in 2015. The ITU is concerned with the global use of the radio spectrum. In general, each State has the sovereign right to regulate its own telecommunication. Control and establishment of means of communication has traditionally been a State affair. However, for an efficient and practical use of (international) communication, it was thought useful to establish rules on systems and procedures that all States agreed on and complied with.[4] Consequently, an international organisation was founded on the principle of international cooperation between States and the private sector.[5]

Following article 1 of the Constitution of the ITU,[6] the purposes of the ITU are, *inter alia*, as follows:

- to maintain and extend international cooperation among all Member States for the improvement and rational use of telecommunications;
- to promote the development of technical facilities and their most efficient operation with a view to improving the efficiency of telecommunication services, increasing their usefulness; and
- to promote at the international level the adoption of a broader approach to the issues of telecommunications in the global information economy and society, by cooperating with other world and regional intergovernmental organisations and those non-governmental organisations concerned with telecommunications.

With these aims in mind, the ITU is responsible for the allocation of bands of the radio-frequency spectrum, the allotment of radio frequencies and the registration of radio-frequency assignments. In that context, its goal is to avoid harmful interference between radio stations of different countries. One of the resolutions of the ITU states that the mission of the ITU Radiocommunication Sector is to ensure the rational, equitable, efficient and economical use of the radio-frequency spectrum by all radiocommunication services, including those using satellite orbits.[7]

---

[4] Francis Lyall, *International Communications, the International Telecommunications Union and the Universal Postal Union* (Ashgate 2011), pp. 3–4.

[5] Constitution of the International Telecommunication Union 1992, article 2.

[6] Constitution of the International Telecommunication Union 1992, article 1.

[7] Radio Regulations 1995, Annex to Resolution 71, Background on the strategic plan for the Union for 2016–2019.

Filing Procedure Within the ITU Framework

The procedure for coordination of the use of frequencies is specified in the ITU Radio Regulations (hereinafter 'RR'). It consists, in short, of the following steps:

- advance publication information (API)[8];
- coordination with other States[9];
- notification and recording of the specific frequencies in the Master International Frequency Register (MIFR).

The latter contains all frequency usage notified to the ITU. If a governmental or non-governmental entity is planning on launching a satellite, the MIFR should be consulted first before selecting a frequency. In this sense, it is very important for administrations to notify the frequency assignments, as the MIFR constitutes the international 'database' of all frequency usage.

The notification procedure is laid down in specific articles of the ITU RR. Normally, for geostationary and non-geostationary satellites, an administration has to send the API to the ITU not earlier than 7 years and not later than 2 years before the planned date of bringing into use of the network or system.[10] The planned date of bringing into use of the network or system is therefore an important date in the filing procedure as a whole. Any frequency assignment *not* brought into use within the required period shall be cancelled by the ITU after having informed the administration at least 3 months before the expiry of this period.[11] This is in line with the goals of the ITU, to assure efficient use and equitable access. However, for specific space objects, there are shorter processes (depending on the frequency band that is used) in place.[12]

Recent Developments and Issues

The way the filing procedure as described above has to be followed does not allow for a lot of flexibility. When an entity plans to launch one satellite, which will only use one specific frequency, this should normally not have problematic consequences. However, in regard to satellite constellations, the question arises how to cope with the API. A satellite constellation consists of multiple satellites, often using several frequencies. These satellites are often not launched at the same time. This raises questions in terms of when to notify the usage of frequencies. Does an

---

[8] Radio Regulations 1995, Article 9, Section I.

[9] Radio Regulations 1995, Article 9, Section II.

[10] Ibid. under n. 9 and 10.

[11] Radio Regulations 1995, Article 11.44.

[12] United Nations Office for Outer Space Affairs and the International Telecommunction Union UNCOPUOS (Legal Subcommittee) Fifty-fourth session 13-24 April 2015 'Guidance on Space Object Registration and Frequency Management for Small and Very Small Satellites' (13 April 2015) A/AC.105/C.2/2015/CRP.17.

administration file all frequencies at once, for all planned satellites? If yes, when does the network of satellites qualify as being brought into use? Does this start when the first satellite is brought into use? This could be problematic, as the first satellite of a constellation could be brought into use years before the others. In this sense, the filing procedure in regard to satellite constellations could lead to warehousing of frequencies, as private actors will be driven by economic considerations and will want to secure multiple frequency filings. Spectrum warehousing in its turn would not constitute as an efficient use of frequency spectrum. Several players from the industry, who argue for more clarity in the RR, also address this issue. They asked for clear regulations regarding the BIU procedure, for example through adding additional milestones. The issue concerning the BIU procedure and satellite constellations was recognized by the ITU at the WRC-15. At this WRC, the ITU-R was invited to examine provisions requiring additional milestones beyond normal notification and the bringing into use procedure.[13]

1.2.2 Space Law

The Corpus Iuris Spatialis

The United Nations treaties on the use of outer space provide for a global legal framework related to all space activities. Every satellite that is launched, whether it is a large GEO satellite or a satellite that is part of a constellation of small LEO satellites, falls within the definition of a 'space object'. Following article 1 of the Liability Convention, a space object includes component parts of a space object, as well as its launch vehicle and parts thereof. The Outer Space Treaty, 'the constitution' of outer space, celebrated its 50th anniversary on 27 January 2017. The Treaty stipulates in article VI that States Parties to the Treaty shall bear international responsibility for national activities in outer space, carried out by both governmental and non-governmental entities. Part of this article is the regime of authorization and continuing supervision by the appropriate State Party to the Treaty.

Article I of the Outer Space Treaty stipulates that the exploration and use of outer space shall be carried out for the benefit and in the interests of all countries, irrespective of their degree of economic or scientific development, and shall be the province of all mankind. Throughout the Outer Space Treaty (and the other treaties on the use of outer space), the principle of cooperation and equal use for all States is referred to. 'The province of all mankind' phrase of article I is a phrase that is used in relation to other earthly resources as well, for example the high seas and the deep sea bed. It relates to the 'common heritage' of all mankind and highlights the goal of the Outer Space Treaty that the interest of all mankind should be taken into consideration in the use and exploration of outer space, this without any

---

[13]Yvon Henri, 'Regulatory challenges for small satellites and new satellite constellations' (ITU International Satellite Symposium 2017, Bariloche, 29-31 May 2017).

discrimination to non-spacefaring countries.[14] A similar wording can be discerned in the ITU framework, as both legal regimes are established on the principles of equal use and international cooperation.

Recent Developments and Issues

As we have seen over the last years, the surge of private actors in outer space has resulted in changes in the way we view the regulatory and legal framework of the space treaties. In addition, it has resulted in several challenges for States Parties to accommodate and facilitate new developments within the current regulatory and legal framework.

Whereas the use and exploration of outer space used to be an area limited to the two Cold War powers the Soviet Union and the United States, many other States and private actors have emerged in this field. Despite the fact that the Outer Space Treaty is the core of the United Nations treaties on the use and exploration of outer space and the fact that it is a very important treaty, it is also a fact that it was drafted over 50 years ago in a very different time and age. In this context, many authors have argued that the Outer Space Treaty is in need of an update as it does not address current issues. Although it is definitely true that specific developments were not foreseen (who knew in the 1960s that we would have the technology to land on asteroids or who would have dreamed of having an international space station up in outer space?), this does not mean that the Outer Space Treaty is outdated. It is still a very relevant framework, as it is a set of rules that allows for a lot of flexibility for States Parties.

Because the Outer Space Treaty focuses on States Parties only (who, in their turn, are responsible for the activities of non-governmental entities), the privatisation of outer space and recent technological developments ask for a different, more flexible approach to the current legal space framework.

1.2.3   The Relation Between the Legal Regimes

Although both legal regimes were traditionally established from different backgrounds (the UN treaties provide general provisions on the use and exploration of outer space and is focused on State responsibility for non-governmental and governmental activities, whereas the ITU framework focuses on the use of telecommunications and also addresses private actors), these regimes are intrinsically linked to each other. As pointed out in the previous paragraphs, both regimes focus on the equal use of outer space and its resources, on harmonisation and on international

---

[14] Stephan Hobe, 'Article I of the Outer Space Treaty' in Prof. Dr. Stephan Hobe, Dr. Bernard Schmidt-Tedd and Prof. Dr. Kai-Uwe Schrogl (eds) *Cologne Commentaries on Space Law* (Wolters Kluwer 2013).

cooperation. The Outer Space Treaty focuses on liability and responsibility of States for non-governmental and governmental entities. It is addressed to States. In a sense, the ITU RR also address States, but they also focus on private actors. This distinction is made even more clear when one considers that the Outer Space Treaty was established in 1967 and has not been amended, whereas the ITU adopts multiple resolutions at each World Radio Conference. Thus, the framework of the ITU allows for more flexible adaptation of the rules and regulations, whereas the Outer Space Treaty constitutes a more robust framework.

When an entity plans on launching a satellite, both legal regimes come into play. Many governments have separated these regimes and have separate authorities concerned with the authorisation of space activities and those concerned with frequency filings. However, to launch a satellite and to use outer space, international coordination in terms of frequency allocation is needed. It is in this light that the author argues that with the surge of satellite constellations and large satellite infrastructures, both regimes will become more and more interrelated. In addition, as the legal framework of the UN Treaties on the use and exploration of outer space allows for a lot of flexibility for States Parties, clear rules and regulations need to be in place to facilitate the use and exploration of outer space through satellite constellations.

1.3 Current Issues

1.3.1 Scarcity of Frequencies and the Role of Governments

The frequency spectrum is perceived as a scarce or even finite resource. The surge of technological advancement and developments can result in not only congestion of specific orbits but also a scarcity in specific frequency bands. When more and more satellites are launched, the demand for specific frequencies will rise. As is the case for many objects and resources, when the demand for spectrum increases, the scarcity of frequencies will increase as well.

In the context of spectrum used for broadcasting and mobile applications, it is usually the prerogative of governments to decide how to divide this spectrum. Most governments state that it is their duty to allocate radio spectrum in such a way that it is of overall benefit to the economic, social and cultural interests that are attached to frequency use. In addition, to prevent harmful interference, spectrum needs to be managed, either through governments or international bodies.[15]

For filing procedures, however, administrations have to adhere to the rules and regulations of the ITU. The API procedure is a strict, formal procedure, and it does not allow for many deviations.

---

[15] See for example the Netherlands Radio Spectrum Policy Memorandum (2016).

1.3.2   Spectrum Warehousing

Because frequencies in specific bands are scarce, the filing procedure is strict and the RR are not always clear on the procedure related to satellite constellations, this could possibly lead to spectrum warehousing. The driving actors behind the rise of satellite constellations are private actors, companies that are, in majority, driven by economic considerations. Frequencies are scarce and filings can be seen as very valuable assets. If regulations regarding the filing procedure of satellite constellations are unclear, this could lead to spectrum speculation. It is evident that such developments are not in line with the idea of the efficient use of frequency spectrum.

Satellite constellations often use non-geostationary orbits. It is unclear when frequencies of a constellation are brought into use. If a constellation consists of, for example, 200 satellites, which will not be launched at the same time, these satellites will have different dates of bringing frequencies into use. In that context, the question arises when the network or system is being brought into use. Is this from the moment the first satellite is launched? Does that mean that the other, not yet launched satellites have the same date of being brought into use? If so, this could mean that entities could file for a significant amount of frequencies that might be brought into use later than 7 years after the API (or even not at all). This is problematic because all frequency usage is notified to the ITU and recorded in the MIFR. The MIFR functions as the international database of all frequency usage and consulted by all administrations before notifying specific frequency assignments. In this sense, spectrum speculation could lead to 'paper filings' and a very inefficient use of spectrum.

Following article 11.44 of the ITU RR, a frequency assignment to a geostationary system is perceived as being brought into use when a space station in the geostationary orbit, with the capability of transmitting or receiving that frequency assignment, has been deployed and maintained at the notified orbital position for a continuous period of 90 days. The current practice at the ITU for non-geostationary systems (as are most satellite constellations) is similar to that of geostationary systems. A frequency assignment is brought into use when there is one non-geostationary satellite at one orbital plane, capable to transmit or receive frequency assignment for a continuous period of 90 days.[16] This also follows from the Rules of Procedure belonging to article 11.44 of the ITU RR.

---

[16]Yvon Henri, 'BR Director's Report to WRC-15, NGSO Issues' (Workshop on the Efficient Use of the Orbit Spectrum Resource, Danang, 2015).

1.4   Considerations on the Current Issues

As was explained in Sect. 1.2.2, the UN Treaties on the use and exploration of outer space focus on States Parties. The Treaties are established on the idea of liability and State responsibility for non-governmental and governmental activities in outer space. Although scholars have argued that the Outer Space Treaty, the core of the UN Treaties on the use and exploration of outer space, is outdated because it was drafted in the Cold War period, it provides for a robust framework. Because it has not been amended since it was drafted in 1967, several soft law instruments have been established. States Parties are in a way driven to find flexible ways to react to current and future developments, within the framework of the UN Treaties on the use and exploration of outer space. For example, the United Kingdom is looking into the traffic light approach for small satellites, which is a proposal to simplify the licensing procedure for small satellites. Other examples can be found in the Hague Space Resources Governance Group, a group that was established in the light of recent developments concerning space mining and the use of space resources. This group examines obligations regarding on-orbit operations and space resource rights. An example of a soft law approach to current and future development is the IADC Space Debris Mitigation Guidelines, which are adopted by several States.

All these examples show that within the current framework of the Outer Space Treaty and the other Treaties related to the use and exploration of outer space, States and international groups or organisations have the flexibility to adapt to new developments. The ability to adapt to new developments will also be necessary in light of the surge of satellite constellations and possible spectrum warehousing. Governments will have to be stricter when it comes to applications for frequency spectrum. The ITU framework related to satellite communications does not allow for government involvement in a way that might be possible in relation to broadband and mobile spectrum. Governments can, however, set up rules for private actors to show that they are planning on bringing their satellite networks into use before a specific date, for example by emphasising the use of milestones. The use of milestones is a current practice in space contracts; space companies are familiar with and used to having milestones.

There is also a big role for the ITU. It recognized the issues with frequency assignments of satellite constellations. It suggested for the WRC-19 to redefine the system of bringing into use. To avoid spectrum warehousing, clear rules and regulations need to be established, and the ITU WRC is the ideal platform to do this. If the ITU adopts clear resolutions on bringing into use satellite constellations and provide governments and non-governmental entities with more guidance on this topic, it will be easier for these entities to adapt to the current legal framework. Uncertainties in regulations can result in entities finding loopholes, or in this case speculating with spectrum. In this context, clear resolutions adopted by the ITU will give governments an instrument to use in their relation with those private actors that might be suspected of spectrum warehousing.

1.5   Conclusion

The UN Treaties on the use and exploration of outer space provide a robust framework based on the liability and responsibility of the State for governmental and non-governmental activities. This basis is needed for legal security for both States and private actors. Within this framework, the regulatory side of frequency assignments is developed by the ITU. Both legal regimes focus on the equal and equitable use of outer space and its resources, and this focus should be leading in their approach to new developments such as the rise of satellite constellations.

With the surge of satellite constellations, more and more satellites will be launched in the future. Unclear provisions in the RR could lead to speculation on spectrum, and possibly also to spectrum warehousing. As both the UN Treaties on the use and exploration of outer space and the ITU Convention and RR focus on the equal use of outer space and its resources, both regimes should address the current issues at hand and avoid spectrum warehousing.

Here lies a role for the administrations applying for frequency allocations on behalf of governmental and non-governmental entities. These administrations have the possibility to pose stricter rules on the information that is provided to them by these entities. However, more important in this regard is the role of the ITU. If, for example, more milestones are incorporated in relation to the bringing into use procedure of satellite constellations, enforcement will be easier from the point of view of administrations responsible for the filing of frequencies on behalf of both private and non-private actors.

The ITU already recognized the issues related to the filing of frequencies for non-geostationary satellites in satellite constellations. It addressed these issues at the WRC-15, and it is also an agenda item for WRC-19. Because the ITU legal framework is quite adaptable—each WRC new resolutions are adopted—it is of utmost importance to provide clear rules and regulations on satellite constellations and frequency assignments. This is even more important because at the WRCs, the majority of all States are present. Traditionally, technological developments in outer space have been developing quicker than laws and regulations, but in this case the global community has the opportunity to adopt resolutions that can accommodate these technological developments in time.

Margaux Morssink graduated from the University of Amsterdam with a Master's Degree in International Law. In addition, she obtained a Master's Degree in Space Activities and Telecommunications Law from the Institut du Droit de l'Espace et des Télécommunications (Université Paris-Sud). She currently works at Agentschap Telecom, the Dutch Radio Communications Agency, at the legal department. Agentschap Telecom is responsible for obtaining and allocating frequency spectrum and monitoring its use. It is the notifying administration to the ITU, and it regulates Dutch space activities. margaux.morssink@gmail.com.

Chapter 2
The Principle of Equitable Access in the Age of Mega-Constellations

Matteo Cappella

Abstract The principle of equitable access was introduced by the International Telecommunications Union to ensure that every nation, spacefaring or not, would have the possibility, at any time, to have access to space and to the necessary spectrum to communicate to and from satellites, without creating or receiving interferences to and from others. The principle applies specifically to the orbital slots and spectrum allocation procedures for the geosynchronous orbits belt, from where broadcasting has been conducted since the 1960s. Such principle does not have, instead, a direct enforcement in the allocation procedures for satellites in non-geosynchronous orbits. Given today's increasing number of satellites proposed and launched, in particular as part of (mega-)constellations, and given the increasing concerns related to overcrowded low Earth orbits, this should be the right time to raise the issue on starting enforcing the principle in all orbits, before non-yet-spacefaring nations find themselves incredibly thwarted in launching one or more satellites, let alone fairly competing with space powers and spacefaring corporations. Opening the debate might be worth the effort, even just as a reminder that space is province of all mankind.

2.1 Introduction

The principle of equitable access is the assumption according to which each nation, spacefaring or not, should have the possibility, at any time, to have access to space and to the necessary spectrum to communicate to and from satellites, without posing or receiving interferences to others. The principle is part of Article 44 of the International Telecommunications Union's Convention[1] and applies specifically to the geosynchronous orbits (GSO) belt, the belt at circa 35,700 km of height above Earth's equator, from which broadcasting has been conducted since the launch of the first telecommunication satellite in the 1960s.

---

[1] Constitution of the International Telecommunication Union (adopted in 1992) art 44.

M. Cappella (✉)
European Space Policy Institute (ESPI), Vienna, Austria

© Springer Nature Switzerland AG 2019   11

A. Froehlich (ed.), *Legal Aspects Around Satellite Constellations*, Studies in Space Policy 19, https://doi.org/10.1007/978-3-030-06028-2_2

Today, non-GSO satellites—therefore all those from the Karman line up to the GSO belt, belt excluded—are still not governed by equitable access principle, but rather the allocation to these orbits and related spectrum is conducted on a first come, first served basis. Given the increasing number of satellites proposed and launched, in particular as part of non-GSO (mega-)constellations, and given the increasing concerns related to overcrowded low Earth orbits (LEO), this should be the right time to raise the issue to start enforcing the principle of equitable access even to non-GSO, before non-yet-spacefaring nations find themselves unable or incredibly thwarted to launch in LEO.

This essay tries to identify the right arguments to raise the problem in future relevant fora. Section 2.2 will describe how the current allocation mechanism has been developed and envisioned as GSO centred. Section 2.3 will shift the focus on mega-constellations, the issues to them connected, and why, legitimately, the principle of equitable access has not been the main concern so far, but there's room for raising the voice right now. Section 2.4 will try to find good, justifiable reasons on why this might be a false issue, while Sect. 2.5 and the Conclusion will, instead, show why such concern is indeed worth exploring.

2.2   The Allocation of Orbital Slots and Spectrum

2.2.1   The Function of the International Telecommunications Union

The International Telecommunications Union (ITU, the Union), born International *Telegraph* Union in 1865, is one of the oldest intergovernmental organisations. Created with the purpose of overseeing the implementation of the International Telegraph Convention, the Union obtained its current name and structure in 1934, being consequently integrated in the United Nations (UN) structures in 1947.[2]

Among its functions, the Union deals with the management, regulation and allocation of radio frequency worldwide by maintaining and updating the Master International Frequency Register; publishing updated Radio Regulations (RR) with technical requirements for stations, receivers and so on; dividing the physically available spectrum among categories and usages; and doing so as follow-ups of the World Radiocommunications Conference[3] (WRC), which takes place every 3–4 years and in which member states, international organisations and various stakeholders such as telecommunication companies are invited to contribute and find arrangements for transmitting via radio frequencies.

---

[2] 'Overview of ITU's History' (*ITU*) https://www.itu.int/en/history/Pages/ITUsHistory.aspx accessed 02 September 2018.

[3] Before 1995, the World Administrative Radio Conference, WARC, was held instead of the WRC.

As for satellites (defined as space stations by the ITU), the Union not only divides and allocates spectrum accordingly with usages and functions but also assigns the orbital slots in which the allocated frequencies can be used.[4] The underlining reason for the RR is, therefore, that of avoiding interferences and using available spectrum and orbital slots in the most efficient way.

In 1959, during the dawn of the first Space Race, the Union inserted the first definitions of space-related activities in its RR, starting to allocate and manage orbits and spectrum in the following years.[5] Aligned with the principle of efficient allocation, the Union used to distribute orbital slots and spectrum in a *first come, first served* basis, clearing the requests filed by national administrations, provided that the intended satellite network[6] would have respected all the existing RR and not interfered with other satellites or terrestrial systems and networks.[7]

2.2.2 The Principle of Equitable Access

As reported by Audrey L. Allison in *The ITU and Managing Satellite Orbital and Spectrum Resources in the 21st Century*,[8] after the ITU began allocating orbital slots and spectrum, a concern started emerging on the first come, first served approach. Realistically enough, in fact, such approach would have soon deprived not-yet-spacefaring nations from ever orbiting a satellite in the commercially precious GSO belt, given that by the time such nations would have developed the right capabilities to build and launch satellites, the GSO would have already been both overfiled at the ITU and overcrowded in space.

With the 1971 World Administrative Radio Conference (WARC) for Space Telecommunication, the Union took notice of this concerns, and, on the conference's final acts, the delegates stressed out that spectrum and GSO were limited natural resources and that every country had the right to access them[9]—right also indicated by the then 4-year-old Outer Space Treaty[10] (OST), which had entered into force in 1967.

---

[4] Constitution of the International Telecommunication Union (adopted 1992) art 1.

[5] Audrey L. Allison, *The ITU and Managing Satellite Orbital and Spectrum Resources in the 21st Century* (Springer Briefs in Space Development, Springer, 2014) 13.

[6] Defined by the ITU as existing network of satellite plus one or more ground stations.

[7] Audrey L. Allison, *The ITU and Managing Satellite Orbital and Spectrum Resources in the 21st Century* (Springer Briefs in Space Development, Springer, 2014) 14.

[8] Ibid.

[9] WARC, 'Resolution No. Spa2–1 relating to the Use by all Countries, with equal Rights, of Frequency Bands for Space Radiocommunication Services' (Geneva 1971).

[10] Treaty on Principles Governing the Activities of States in the Exploration and Use of Outer Space, including the Moon and Other Celestial Bodies (adopted 27 January 1967, entered into force 10 October 1967) (Outer Space Treaty) art I.

Finally, in 1973, the Union's Plenipotentiary Conference addressed the issue by adding, in Article 33 of the Union's convention, that

> In using frequency bands for space radio services Members shall bear in mind that radio frequencies and the geostationary satellite orbit are limited natural resources, that they must be used efficiently and economically so that countries or groups of countries may have equitable access to both in conformity with the provisions of the Radio Regulations according to their needs and the technical facilities at their disposal.[11]

This provision, which introduced the principle of equitable access, started a period of debates and turmoil on how or whether to enforce such principle. Noteworthy, as an example of this period, was the Bogotá Declaration of 1976,[12] with which eight equatorial countries[13] declared sovereignty on the part of the GSO belt directly above their national territory, in clear violation of the OST.

The whole process paused, not really ended, in 1988, when ITU delegates reached a compromise to redefine the allocation process for geostationary orbits.[14]

2.2.3   The Current Allocation Mechanism

The process of allocating orbits and related spectrum starts with the RR, constantly updated to reflect the need of states and industries, as well as to mirror contemporary telecommunication trends, which would show, for instance, the predominance of the use of a frequency band over others or that of a specific telecommunication medium over others.

Regarding the space-related spectrum and orbits, the RR divide which frequency can be used by which type[15] of satellite and orbit, with non-GSO satellites required to regulate the power of their transmissions[16] and to ensure to not interfere with other GSO satellites while the transmissions of the former are on the crossline with those of the latter.

As for satellites stationing in GSO, instead, since the mentioned 1988 redefinition of the allocation process, today a dual, parallel mechanism exists, reflecting the principles of efficiency and of equitable access, discussed by Article 44 of the ITU Convention (the once Article 33). The mechanism, in practice, splits the GSO

---

[11] ITU Plenipotentiary Conference, 'International Telecommunication Convention' (Malaga-Torremolinos 1973) art 33.

[12] Declaration of the First Meeting of Equatorial Countries (adopted 03 December 1976) (Bogotá Declaration).

[13] Brazil, Colombia, Congo, Ecuador, Indonesia, Kenya, Uganda, and Zaire.

[14] Audrey L. Allison, *The ITU and Managing Satellite Orbital and Spectrum Resources in the 21st Century* (Springer Briefs in Space Development, Springer, 2014) 15.

[15] As for Scientific purposes, Deep Space missions, Earth Observation, Telecommunications, etc.

[16] Timur Kadyrov, 'Equivalent Power Flux Density Limits (EPFD)' (*ITU*, 2016) https://www.itu.int/en/ITU-R/space/WRS16space/WRS-16_EPFD.pdf accessed 02 September 2018.

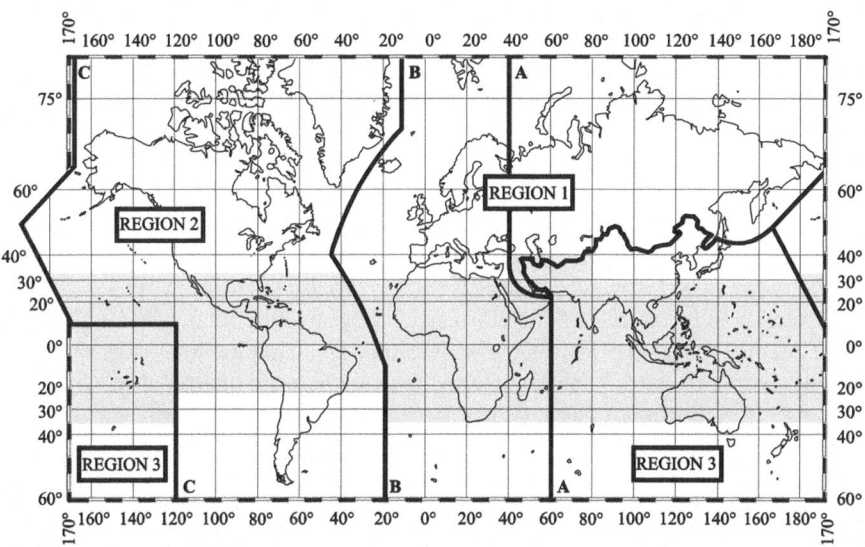

Fig. 2.1 World regions in ITU radio regulations. Source: ITU

allocation procedures into two groups: non-planned (or coordinated) allocation, and planned allocation.

The planned procedures pursue the principle of equitable access and are based on the planning and division of spectrum and orbits along the three regions that the ITU has divided the world into (Fig. 2.1), furthermore reserving bandwidth and GSO slots to each country, ensuring in such way the enforcing of the principle.[17]

The coordinated allocation, instead, pursues the principle of efficient allocation and utilisation, and it is based on the first come, first served approach—provided that every filing is accompanied by coordination efforts with all the administrations and potential stakeholders actively or passively impacted by the new satellite, including competitors of both manufacturer and operator of the satellite.[18]

Finally, and relevantly enough for the emergence of low Earth orbits satellite mega-constellations, it is necessary to mention that non-GSO and related frequency spectrum are allocated today only with the coordinated procedures, therefore solely on the first come, first served basis.[19]

---

[17] Audrey L. Allison, *The ITU and Managing Satellite Orbital and Spectrum Resources in the 21st Century* (Springer Briefs in Space Development, Springer, 2014) 17.

[18] International Telecommunications Union, *Radio Regulations* (edn. 2016, vol. 1) art 9.

[19] Yvon Henri, 'Orbit/Spectrum International Regulatory Framework' (*ITU*, 2016) https://www.itu.int/en/ITU-R/space/Presentations/Orbit_Spectrum%20International%20Regulatory%20Framework_Henri.pdf accessed 02 September 2018.

2.2.4   A GSO-Centred Mechanism

As one can notice, the overall mechanism appears undoubtfully GSO centred. The features and limitations of non-GSO satellites are specifically envisioned not to harm GSO transmissions, while, as seen, the current allocation procedures are a result of clashes and debates exclusively driven by the use of geostationary orbits (GEO), with the principle of equitable access having been fought over the specific access to GSO rather than LEO or medium Earth orbits (MEO).

One ought to acknowledge that such GSO-centricity has made sense so far. From a physical standpoint, the GSO belt is particularly spatially limited and at the same time peculiarly optimal for telecommunication applications. Moreover, it makes physical sense to restrict non-GSO usage to avoid harmful interferences with GSO stations, given that non-GSO satellites are stationed in between Earth and GSO and therefore every communication to and from the higher orbits is necessarily transmitted through lower orbits. Adding the fact that GSO satellites are a very wealthy business, it makes sense for a nation to focus its attention on gaining access, present or future, to such orbits. In 2016, the satellite service market, and specifically for Broadcast and Fixed Satellite Services, amounted to USD 121.1 billion[20]: doubtlessly no nation could desire to be excluded from it.

Despite this, looking at the recent surge in LEO mega-constellation proposals, the GSO-centricity of the allocation process for spectrum and orbital slots could start not being justifiable anymore. As constellations of hundreds or thousands of small satellites are going to swarm non-GSO orbits, directly challenging the business case for GSO networks, it might be time to give voice to concerns over the regulatory procedures to access to LEO and MEO, especially on the lack of an enforcement of the equitable access principle. With a global slowdown in orders for GEO satellites,[21] and a surge in smallsats destined to LEO,[22] the equitable access principle might otherwise end being applied for just a small minority of all the satellites currently orbiting the planet, and potentially not even the most relevant for the future space economy.

---

[20] Bryce Space and Technology, 'State of the Satellite Industry Report' (*SIA*, 2017) https://www.sia.org/wp-content/uploads/2017/07/SIA-SSIR-2017.pdf accessed 02 September 2018.

[21] Kendall Russell, 'How Boeing is Coping with the Slowdown in GEO Orders' (*Via Satellite*, 18 July 2017) https://www.satellitetoday.com/innovation/2017/07/18/boeing-exec-surviving-slowdown-geo/ accessed 02 September 2018.

[22] Jeffrey Hill, 'Report: Six-Times More Smallsats Will Launch in the Next Decade' (*Via Satellite*, 06 August 2018) https://www.satellitetoday.com/launch/2018/08/06/report-six-times-more-smallsats-will-launch-in-the-next-decade/ accessed 02 September 2018.

2.3 The Issues with Equitable Access and Constellations

2.3.1 The Scale of the New Mega-Constellations Proposals

When the WARC-92 approved the use of frequency for the emerging mobile satellite systems called Little LEO and Big LEO—an approval that was highly lobbied for by the corporation Motorola and its child company Iridium[23]—the idea on such systems was that they were a rather eccentric hubristic proposal, mostly bound to fail, at least just given the amount of political and commercial actors directly impacted and potentially disrupted by the concept.[24]

Following such decision, in 1996, the ITU held a policy forum to discuss the regulatory implications on the topic, acknowledging the existence of 16 proposed constellations, for a total of 1315 satellites, of which 840 were part of a single constellation envisioned by Teledesic.[25]

Such amount of satellites was indeed large enough to be both seen as ambitious and discredited as wishful thinking.

Looking at it in 2018, however, the scale of those proposals is far smaller than that of today's constellations: there are currently 58 commercial planned constellations, excluding those of companies and governments operating under the radar; the envisioned systems range from dozens of CubeSats to 5000 (sometimes suggested 13000) satellites; and the conservative total figure amounts to ~15,000 satellites to be manufactured and launched during the 2020s, targeting exclusively non-GSO, mostly LEO.[26,27]

2.3.2 Other Concerns Overshadow a Lack of Equitable Access

These numbers are indeed impressive, and attention has been drawn to the many issues, challenges, and even threats that such amount of satellites could pose.

Among the concerns, many of them remain GSO centred, as there are justifiable reasons to doubt the capabilities of operators to be able to ensure that every satellite will perfectly avoid interference with GSO satellites. A similar concern is related to the fact that, as smallsats and CubeSats are launched in large batches, up to 104 so

---

[23] John Bloom, *Eccentric Orbits: The Iridium Story* (Groove Atlantic, 2016) 103–104.

[24] Ibid, 116–118, 125–135.

[25] 'World Telecommunication Policy Forum Fact Sheet' (*ITU*, 1996) https://www.itu.int/newsarchive/wtpf96/fact.html accessed 02 September 2018.

[26] 'NewSpace Constellations' (*NewSpace Index*, 18 August 2018) https://www.newspace.im/index.html accessed 02 September 2018.

[27] Sebastien Moranta and Matteo Cappella, 'The Digital Transformation of the Satcom Sector: between Opportunities and Challenges' in Laurence Nardon (ed.) *European Space Programs and the Digital Challenge* (Études de l'Ifri, Ifri, November 2017) 59.

far,[28] there might be the risk of *rogue* satellites, either not transmitting as licensed or directly launched without any ITU license, as it has recently happened,[29] therefore further increasing risks of unchecked transmissions and eventual interferences.

Stakeholders are also worried about LEO overcrowding and the capability of operators to actively avoid space debris or other satellites.[30] CubeSats, for instance, are rarely developed with propulsion systems, meaning that every CubeSat launched into orbit is totally vulnerable to debris of any kind. In addition to this, having more and more satellites into orbit simply increases the probability of collisions, which would cause further debris and space junk. Finally, and more in general, the overall long-term sustainability of such approach to space is doubted,[31] as non-GSO constellations allows for/requires a constant turnover and replacement of satellites, therefore increasing the amount of space junk.

Given the dimension of these concerns (and the actors that express them), the lack of voices raised on the issue of equitable access could at first be considered even reasonable: the dimension of the other concerns and the novelty of the trend never alimented a proper debate.

In fact, the issue is pretty novel, as the first constellations of CubeSats started appearing only in 2013,[32] while Iridium and Globalstar, the two Big LEO systems still operating after the 1990s, had never shown the interest to increase the dimension of their constellations. Given also that the last WRC—held in late 2015—was contemporary to the sudden burst of proposals for non-GSO mega-constellations, it is understandable that no discussion has been funnelled yet.

2.3.3   There Is Enough Room to Raise the Issue

Even in the light of these justifications, bearing in mind that there is no equitable-access–inspired procedure for non-GSO allocation, the launch of more than 15,000 satellites might indeed represent an obstacle and impediment to orbiting and

[28] Michael Safi, 'India launches record-breaking 104 satellites from single rocket' (*The Guardian*, 15 February 2017) https://www.theguardian.com/science/2017/feb/15/india-launches-record-breaking-104-satellites-from-single-rocket accessed 02 September 2018.

[29] Mark Harris, 'FCC Accuses Stealthy Startup of Launching Rogue Satellites' (*IEEE Spectrum*, 09 March 2018) https://spectrum.ieee.org/tech-talk/aerospace/satellites/fcc-accuses-stealthy-startup-of-launching-rogue-satellites accessed 02 September 2018.

[30] Jon Brodkin, 'SpaceX and OneWeb broadband satellites raise fears about space debris' (*Ars Technica*, 04 October 2017) https://arstechnica.com/information-technology/2017/10/spacex-and-oneweb-broadband-satellites-raise-fears-about-space-debris/ accessed 02 September 2018.

[31] Benjamin Bastida Virgili, Juan Carlos Dolado, Hugh G Lewis, et al, 'Risk to space sustainability from large constellations of satellites' in *Acta Astronautica* (no 126, 2016) 154–162.

[32] 'Planet Labs Reveals First Images from Space; Announces Plans to Launch Fleet of Satellites to Understand the Changing Planet' (*BusinessWire*, 26 June 2013) https://www.businesswire.com/news/home/20130626005421/en/Planet-Labs-Reveals-Images-Space-Announces-Plans accessed 02 September 2018.

operating spacecraft in a near future. The disparity of satellites owned in non-GSO would peak drastically, even more than the current situation, with the usual handful of countries owning the quasi totality of satellites and therefore unilaterally benefiting from those orbits and spectrum, which supposedly are open to everyone.

Today's not-yet-spacefaring or still-not-interested nations would end up impeded or incredibly slowed down by the very existence of foreign mega-constellations, even ending up being dependent on, and directly targeted by, the services provided by these.

Such prospect needs also to be analysed in the light of the already discussed serious and informed concerns for space debris, space sustainability and the overcrowding of LEO, together with the ITU acknowledgement, since 1998, that all the orbits—and not GSO only—are limited natural resources[33]: in fact, provided that there is a recognised scarcity of orbits and spectrum, and provided that there are indeed concerns on the long-term sustainability of systems as the mega-constellations, the lack of any discussion on equitable access for non-GSO is not only worrying but also potentially misaligned and conflicting with Articles I and II of the Outer Space Treaty.[34]

2.4 Is the Lack of an Equitable-Access–Inspired Mechanism for Non-GSO a Real Issue?

Given the nature of the principle and the lack of concerned actors, some elements can actually be considered to dismiss such issue as a non-issue.

2.4.1 The Special Status of Geosynchronous Orbits

When considering whether applying the principle of equitable access to non-GSO, one should first remember that GSO might be far more precious.

First of all, a satellite is incredibly worthy to nations when along the GEO belt, given its stationarity over specific locations. For such reason, the equatorial countries signed the Bogotá declaration, and for the same reason today the planned procedures exist.

Moreover, looking at the scarcity of available space, compared to LEO and even more to MEO, the GEO belt occupies a very small volume of space. Even considering that LEO is getting crowded, the volume that MEO encloses is overall 150 times

---

[33] Until 1998, Art 44 of the ITU Constitution expressed "Members shall bear in mind that radio frequencies and the geostationary-satellite orbit are limited natural resources". After 1998 amendment, Art 44 says "Member States shall bear in mind that radio frequencies and any associated orbits, including the geostationary-satellite orbit, are limited natural resources".

[34] Outer Space Treaty, arts I-II.

larger than GEO[35] and five times less populated,[36] offering a right compromise for eventual latecomers interested in launching any kind of constellations, with plenty of space and acceptable remoteness from the ground compared to GEO.

Thus, not only is GSO more precious in terms of operational functionality, but non-GSO is comprehensively less crowded and with a wider volume available.

2.4.2   History Repeats Itself

Such concern could be dismissed by looking at history: recalling the 16 acknowledged proposals for commercial constellations in the 1990s, today only two of them are operative and both went through bankruptcy procedures at the turn of the millennium.[37,38] A third proposal, the Inmarsat-backed ICO, bankrupted before it successfully launched a satellite in 2001, filing again for bankruptcy just three years after, and the constellation never came to life.[39] Eventually, of the 1315 proposed satellites, less than 200 reached orbit, as a painful reminder that one thing is filing for a satellite constellation and one thing is bringing it into use. Some proposals never even began being constructed, suggesting that the filing of the satellite system was solely a way to keep up with the competition and/or pursue orbits and spectrum warehousing with so-called *paper satellites*, a practice that had emerged as a major concern for ITU just during the same decade.[40]

As history shows, it often repeats itself: many constellation proposals have stalled, and some are proceeding at an incredible slow pace,[41] while some were never even pursued after their announcement, as it appears for—among others—the Samsung-proposed constellation.[42] Eventually, the fear of LEO oversaturation

---

[35] Peter B. de Selding, 'Overcrowding Not a Problem in Vast Medium Earth Orbit' (*SpaceNews*, 11 October 2010) https://spacenews.com/overcrowding-not-problem-vast-medium-earth-orbit/ accessed 02 September 2018.

[36] 'UCS Satellite Database' (*Union of Concerned Scientists*, 30 April 2018) https://www.ucsusa.org/nuclear-weapons/space-weapons/satellite-database#.W4LhtugzY2w accessed 02 September 2018.

[37] 'Iridium files chapter 11' (*CNN Money*, 13 August 1999) https://money.cnn.com/1999/08/13/companies/iridium/ accessed 02 September 2018.

[38] 'Globalstar files for bankruptcy' (*BBC News*, 17 February 2002) http://news.bbc.co.uk/2/hi/business/1825743.stm accessed 02 September 2018.

[39] 'With $750M in Debt Coming Due, ICO Files for Bankruptcy Protection' (*SpaceNews*, 29 June 2004) https://spacenews.com/750m-debt-coming-due-ico-files-bankruptcy-protection/ accessed 02 September 2018.

[40] Audrey L. Allison, *The ITU and Managing Satellite Orbital and Spectrum Resources in the 21st Century* (Springer 2014) 26–29.

[41] Caleb Henry, 'Boeing Constellation Stalled, SpaceX Constellation Progressing' (*SpaceNews*, 27 June 2018) https://spacenews.com/boeing-constellation-stalled-spacex-constellation-progressing/ accessed 02 September 2018.

[42] Veronica Magan, 'Samsung Exec Envisions LEO Constellation for Satellite Internet Connectivity' (*Via Satellite*, 18 August 2015) https://www.satellitetoday.com/telecom/2015/08/18/samsung-exec-envisions-leo-constellation-for-satellite-internet-connectivity/ accessed 02 September 2018.

might be exaggerated, and of the quasi 20 thousand proposed satellites, only less than 10% of these might be launched, and these being most likely short-lived CubeSats or smallsats.

2.4.3 The Burden of Further Regulations

Finally, if addressing the issue means speaking about further regulations, one should remember that the allocation for non-GSO satellites do have an enforced orderly mechanism, as explained in Sect. 2.2.3. Each actor has the obligation to coordinate with every potential stakeholder, with the ITU making sure that allocation is done in the most efficient way possible, in a process that lasts up to 7 years.[43] Thus, trying to enforce the principle of equitable access in non-GSO might have an opposite effect: by introducing planned orbital planes and frequencies, further spectrum divisions or a similar additional layer of bureaucracy, it would only increase the already great difficulties that a satellite operator or administration, and in particular an inexpert one, has to overcome in filing, launching and operating satellites in compliant ways.

2.5 The Risks of Dismissing Such Issues

Despite the justifiable ground of the points raised in Sect. 2.4, the dismissal of the concern cannot be carried on lightly.

2.5.1 The Injustice of First Come, First Served

In fact, even though the non-GSO volume is way wider, and acknowledging that MEO in particular is far from overcrowded, it is still necessary to consider that there are easier, cheaper and more long-lasting orbits than others. In fact, different orbital inclinations, further orbital distances or orbital velocities all imply different launch and operational costs, impacting the business case and the utility of constellations accordingly. In addition to this, there's also the Van Allen radiation belt[44] to take into account, especially when dealing with satellites in MEO.

These elements need all to be considered, as, eventually, some architectures won't work or will be too expensive to commercially pursue, meaning that there won't be real freedom of choice for latecomers as all the best orbits have been

---

[43] International Telecommunications Union, *Radio Regulations* (vol. 1, edn. 2016) art 11.44.

[44] 'Radiation: Satellites' Unseen Enemy' (*ESA*, 04 July 2011) http://www.esa.int/Our_Activities/Space_Engineering_Technology/Radiation_satellites_unseen_enemy/ accessed 02 September 2018.

assigned earlier: operating only with a first come, first served approach means that the last to apply will end up with poor conditions to work with. Closer and cheaper orbits will be already crowded, therefore making it more expensive for latecomers to launch constellations. The necessity to operate in harsher orbits (Van Allen belt) will force satellite operators to choose between short-lived satellites, compared to the others in LEO, or more expensive satellites, in order to be more resistant. Latecomers will have to deal with less effective system architectures overall, including reduced coverage both in terms of areas and revisit time, higher latency due to the distance from the ground and so on.

If there is enough volume to host constellations in MEO, it is also true that these constellations are certainly disadvantaged when competing with LEO systems.

2.5.2 The Historical Reason for Equitable Access is Fairness

As for the 1990s' wave of filings for constellations, the different scale of this, compared to the current one, has already been illustrated. Dismissing this problem by simply looking at history will not settle the issue of equitable access, as such principle was introduced in 1973 as a matter of rights and fairness.

Today, with non-GSO constellations, this lack of concern and interventions to ensure equitable access is a dangerous gamble based solely on the hope that not every proposed satellite and constellation will actually be launched. As a consequence of this, nations, companies, universities, etc. might be prevented or incredibly slowed down in accessing what is supposed to be the *province of all mankind*.

2.5.3 The Backfiring of Small Regulation

Finally, as for increasing the bureaucratic difficulties for satellite operators and administrations, if history repeats itself, it is also self-perpetuating: a lack of an equitable-access–driven approach in non-GSO might push (or it has already pushed) countries and firms to file constellations and satellites just to not be left behind, even though such actors might not have plans or resources to work to turn the filings into real space systems. A clear example of this could be those constellations that already seem to be stalled, as well as the Russian-[45] and Chinese-proposed[46] constellations. Once these announcements reach the filing procedures at the ITU, the only result is

---

[45] 'Russia to create orbital Internet satellite cluster by 2025' (*TASS Russian News Agency*, 22 May 2018) http://tass.com/science/1005554 accessed 02 September 2018.

[46] Caleb Henry, 'China Satcom's LEO, HTS projects driven by desire not to fall behind foreign counterparts' (*SpaceNews*, 11 July 2018) https://spacenews.com/china-satcoms-leo-hts-projects-driven-by-desire-not-to-fall-behind-foreign-counterparts/ accessed 02 September 2018.

an overfilling and the consecutive obstruction of an already lengthy bureaucratic process. This can indeed impact all the satellite operators' activities worldwide.

2.6   Conclusion

In 2019, ITU will hold the World Radiocommunication Conference once again, the first one after 2015. In the 92-page-long *Agenda and Relevant Resolutions* towards the WRC-19, the words *equitable access* are mentioned three times, but never to directly address the application procedures for non-GSO satellites. *Constellations* is also mentioned three times, but never to address the great number of satellites that might be launched in the near future.[47]

Now, after having duly reported and justified the reasons for a lack of concern, and after having analysed the elements that instead suggest the necessity to raise the issue, it should be clear that this is, at least, a topic worth being aware of—and open to discuss about.

If a debate takes place, it will be necessary to be careful: overdoing things could suffocate emerging NewSpace endeavours such as the broadband mega-constellations, which could otherwise have incredible impact globally.

Conversely, not initiating a serious discussion might mean allowing the nations with current capabilities—or the most pressing companies—to decide for everyone, even if that decision is simple inaction.

The complexity of the subject does not help, nor does the presence of many interests legitimately involved, but discussion is worth the efforts. Starting a debate, and doing so as inclusively as possible, could be a useful reminder of the OST and its principles, and a positive first step to redefine what equitable access means, in the age of mega-constellations.

Matteo Cappella is currently a Research Intern at the European Space Policy Institute (ESPI), Vienna, working on Space Power studies and the European astropreneurial environment. He holds an MA in International Politics and Economic Relations from the University of Macerata, Italy, with master's thesis on the European Union's Spacepower. His experiences and studies include a previous traineeship at ESPI; a BBA in Economics and Commerce from the Marche Polytechnic University, Italy; and a semester-long exchange at the Vidzeme University of Applied Sciences, Latvia, with a focus on entrepreneurial and cultural studies. matteo.cappella@outlook.com.

---

[47] 'World Radiocommunication Conference 2019 (WRC-19): Agenda and Relevant Resolutions' (*ITU*, 2017) https://www.itu.int/dms_pub/itu-r/oth/14/02/R14020000010001PDFE.pdf accessed 02 September 2018.

Chapter 3
Legal Aspects Relating to Satellite Constellations

Ewan Wright

Abstract Satellite constellations represent a new paradigm of space activities, with vast numbers of smaller satellites operated in the so-called NewSpace industry. With the tide of potential applications and benefits to mankind come risks, notably that of the increased population of low Earth orbit and potential for increases in space debris. This article discusses the legal ramifications of the proposed constellations, including the current legal framework around space debris and collisions, and potential challenges, including lack of legal definitions for space debris and fault making liability difficult to prove. Potential mechanisms for clarification are also discussed.

3.1 Introduction

Since mankind first launched objects into space over 60 years ago, we have induced a growing risk of damage caused by space debris. As we increasingly move into a new age of space utilisation, pioneered by companies rather than nations, and characterised by constellations over single satellites, this article aims to outline the current legal aspects relating to satellite constellations and debris mitigation. Furthermore, future legal mechanisms for promoting the sustainable use of space are discussed.

The introductory section will define and outline constellations and space debris, before explaining why constellations offer a new branch of legal discussion with respect to space debris. The legal background pertaining to space debris and constellations will then be outlined. The latter section will discuss these legalities and how they relate to constellations and space debris, including a case study of the Iridium-Kosmos collision and future considerations.

E. Wright (✉)
University of Sheffield, Sheffield, UK

Satellite Applications Catapult, Harwell, UK

© Springer Nature Switzerland AG 2019

A. Froehlich (ed.), *Legal Aspects Around Satellite Constellations*, Studies in Space Policy 19, https://doi.org/10.1007/978-3-030-06028-2_3

3.1.1 Satellite Constellations

Satellite constellations are systems that involve multiple coordinated satellites, often with the same function and type of satellite. The number of satellites can vary from several to thousands. A handful of constellations exist currently; the largest is currently operated by the US company Planet Labs Inc, comprising 148+ satellites.[1] Future constellations plan for far more satellites than this with figures ranging from 720 planned by OneWeb[2] and 4425 planned by SpaceX's Starlink.[3] These two systems alone would quadruple the number of active satellites[4] and increase the amount of tracked objects by 35%. There are estimated to be over 70 other systems currently in development.[5] Whilst many will not launch, it is clear that this represents a significant change in utilisation of low Earth orbit (LEO). The potential benefits of current and future constellations are great, but legal considerations must explicitly underpin development of such systems to minimise risk to the shared resource of Earth orbit.

3.1.2 Space Debris

This section will outline the current state of space debris to outline the need for future management. For the purposes of this report, only LEO debris is considered as most of the planned constellations plan to utilise this space. LEO is defined by the Inter Agency Space Debris Coordination Committee (hereinafter IADC) as from the Earth surface to an altitude of 2000 km.[6]

There is no legal definition for space debris. The United Nations Committee on the Peaceful Uses of Outer Space (hereinafter UNCOPOUS) 2007 Space Debris Mitigation Guidelines, endorsed by the United Nations (hereinafter UN) General Assembly, defines space debris as "all man-made objects, including fragments and

---

[1] Planet Labs Inc 'Frequently Asked Questions' <https://www.planet.com/faqs/> accessed 28th September 2018.

[2] Federal Communications Commission (2017) 'News Release: FCC Grants OneWeb Access to U.S. Market for its Proposed New Broadband Satellite Constellation' <https://docs.fcc.gov/public/attachments/DOC-345467A1.docx> accessed 21st September 2018.

[3] FCC News Release (2018) 'FCC Authorises SpaceX to Provide Broadband Services via Satellite Constellation' <https://docs.fcc.gov/public/attachments/DOC-345467A1.docx> accessed 21st September 2018.

[4] Union of Concerned Scientists currently tracks 1800 satellites—many of these are in geostationary orbits, so the multiplier for low Earth orbit satellites would be much larger.

[5] Satellite Applications Catapult, 'Small Satellite Market Intelligence Report' (2017) Q4 <https://sa.catapult.org.uk/2018/08/01/small-satellite-market-intelligence/> accessed 21st September 2018.

[6] Inter-Agency Space Debris Coordination Committee (IADC) 'Space Debris Mitigation Guidelines' [2007] < https://www.iadc-online.org/index.cgi?item=docs_pub> accessed 22th September 2018.

elements thereof, in Earth orbit or re-entering the atmosphere, that are non-functional."[7] Since the first launches in the 1960s, there has been a mostly linear growth in the amount of space debris.[8]

There are different ways of categorising debris. Common is by size and therefore whether the object can be tracked. Currently, over 20,000 pieces larger than 10 cm are tracked.[9] Several hundred thousand pieces between 1 and 10 cm and over 100 million pieces smaller than 1 cm are estimated to exist.[10] The purpose of this article is not to convince the reader of the existence or urgency of the problem but to discuss the legal implications it offers.

3.1.3   Constellations and Space Debris

It is helpful to distinguish between constellations and space debris for the purpose of this article. Space debris, under the definition above, does not include active satellites, which the constellations will comprise. However, constellations have a potential to cause an increase in space debris, either through collision or satellite failure. Discussions pertaining to debris are equally applicable to active satellites: to a third party, it makes little difference whether an object in space is active or not—it must still be avoided, and a party must be liable for it. Whilst constellations of thousands of satellites are not in themselves contributing to the space debris problem, they pose new legal questions, which will be discussed here.

The emergence of small satellite constellations has specific implications that differ from single satellite operations of the past. This section will outline three key differentiators of constellations when compared to single satellite systems.

[7] United Nations Office for Outer Space Affairs, 2007, 'Space Debris Mitigation Guidelines of the Committee on the Peaceful Uses of Outer Space' 2007 <www.unoosa.org/pdf/publications/st_space_49E.pdf> accessed 21st September 2018.

[8] National Aeronautics and Space Administration 'Orbital Debris Quarterly News' 2018 22(1) <www.orbitaldebris.jsc.nasa.gov/quarterly-news/pdfs/odqnv22i1.pdf> accessed 21st September 2018.

[9] T.A. Aadithya 'Review Paper on Orbital Debris Mitigation and Removal and a New Model Insight' [2015] IJEEDC <www.researchgate.net/publication/281722682_Review_Paper_On_Orbital_Debris_Mitigation_And_Removal_And_A_New_Model_Insight> accessed 21st September 2018.

[10] IADC 'IADC Assessment Report for 2010' [2013] < https://www.iadc-online.org/Documents/IADC-2011-04,%20IADC%20Annual%20Report%20for%202010.pdf> accessed 27th September 2018.

Quantity

Currently, satellites represent a small proportion of tracked objects in space: less than 10%.[11] However, the proposed constellations represent a great increase in the density of objects on the specific orbits. This increases the tracking complexity and information sharing between operators required to avoid conjunctions between systems and debris.

Increase in Upstart Companies

Whilst not an issue with constellations *per se*, the emergency of many smaller space companies across the world, which may not have large legal resources, means regulations should be clear beyond doubt. A recent case has highlighted the complex legalities of the existing regime.

Swarm Technologies Inc. is a US communication satellite upstart that in January 2018 launched four picosatellites from India without Federal Communications Commission (FCC) approval.[12] The FCC had denied Swarm's application for fear of an increased collision risk due to potential difficulty in tracking the particularly small satellites. The launch company, Indian Space Research Organisation, was presumably unaware of the lack of approval. At the time of writing, the satellites are in orbit without communications to the ground, and Swarm is seeking permission to gather data from the satellites.

The mismatch of national legislation could lead to a *flag of convenience* situation—where companies seek registration wherever it is legislatively beneficial to them.[13] This is particularly relevant to the space industry as the registration, launch vehicle registration, and launch states are often different.

Reduced Redundancy

In addition, new models regarding redundancy are being operated. Satellite design used to consist of large, expensive satellites with large redundancies on heavily tested subsystems. Constellations offer a new way of designing and operating satellites; new system designs now focus on inexpensive designs and plan for a certain

---

[11] Based on 1886 satellites and 20,000 tracked objects, see Union of Concerned Scientists 'UCS Satellite Database' [2018] <https://www.ucsusa.org/nuclear-weapons/space-weapons/satellite-database?_ga=2.235743042.294627627.1538140743-1592752914.1538140743#.W64qTtNKiHt> accessed 27th September 2018.

[12] Mark Harris 'FCC Accuses Stealthy Startup of Launching Rogue Satellites' [2018] IEEE Spectrum <https://spectrum.ieee.org/tech-talk/aerospace/satellites/fcc-accuses-stealthy-startup-of-launching-rogue-satellites> accessed 21st September 2018.

[13] Matthew Kleiman 'Space Law 101: An Introduction to Space Law' [undated] American Bar Association <www.americanbar.org/groups/young_lawyers/publications/the_101_201_practice_series/space_law_101_an_introduction_to_space_law.html> accessed 24th September 2018.

number of failures; redundancy through quantity. If a system is designed allowing a certain number of satellites to fail, could this be interpreted as intentionally causing space debris and contrary to existing guidelines? Whilst no satellite designer envisages designing satellites to fail, failed satellites are by definition space debris—and could have legal consequences as such.

Active Debris Removal

The increase in satellites proposed to be launched has prompted the development of various technologies aimed at physically removing space objects from orbit. For satellite constellations, this may be a requirement in certain orbits if propulsion systems fail. These technologies have not been utilised yet, but they raise legal questions—namely the potential removal of another state's space object.

3.2 Legal Framework

This section will outline the existing legal treaties and guidelines pertaining to satellite constellations and space debris. Space law consists of five international treaties; two of them are particularly relevant: the Outer Space Treaty[14] and the Liability Convention.[15]

In addition to these treaties, the UN has published guidelines on Space Debris Mitigation. Whilst UN General Assembly resolutions and UNCOPOUS documents are not legally binding, they serve as subsidiary means for interpretation and application of the principles outlined in the treaties.[16]

3.2.1 Outer Space Treaty

The first international legislation pertaining to space is the Outer Space Treaty. Formed at a time where space was increasingly used for hostile endeavours, the treaty helped unify the globe in promoting space for peaceful purposes. In this endeavour it was successful, though some articles are ambiguous, and space debris

---

[14] United Nations Office for Outer Space Affairs *Treaty on Principles Governing the Activities of States in the Exploration and Use of Outer Space, Including the Moon and Other Celestial Bodies* (hereinafter Outer Space Treaty) [1966].

[15] United Nations Office for Outer Space Affairs *Convention on the International Liability for Damaged Caused by Space Objects* (hereinafter Liability Convention) [1971].

[16] Ma Xinmin 'The Development of Space Law: Framework, Objectives and Orientations' [2014] UN/China/APSCO Workshop on Space Law <www.unoosa.org/documents/pdf/spacelaw/activities/2014/splaw2014-keynote.pdf> accessed 24th September 2018.

is not mentioned. The Outer Space Treaty forms the foundation for which the later treaties detail and has been ratified by over 100 states.[17]

Article VII makes the first mention of liability: 'Each state … is internationally liable for damage to another State Party to the Treaty or its natural or juridical persons by such object or its component parts on the Earth, in air space or in outer space.' In the context of in-orbit collisions, this article is unworkable: there is no test for liability.

Article IX builds on Article IV and is the longest in the treaty. It deals with the peaceful uses of outer space: avoiding harmful contamination of the environment and harmful interference with another state. It has been criticised as ambiguous and insufficient to impose obligations upon states,[18] including lacking a definition for harm.

The Outer Space Treaty has shortcomings; however, on the road to diplomatic sustainability, one must take small steps. The ambiguities highlighted are mostly clarified in the Liability Convention.

3.2.2   Liability Convention

Published 5 years later than the Outer Space Treaty, the Liability Convention outlines the liability for damaged cause by space objects both on Earth and in space.

Article III addresses damage caused by one space object to another space object and declares that the responsibility lies with the party at fault:

> In the event of damage being caused elsewhere than on the surface of the earth to a space object of one launching State or to persons or property on board such a space object by a space object of another launching state, the latter shall be liable only if the damage is due to its fault or the fault of persons for whom it is responsible.

The convention differentiates absolute liability, in the case of objects colliding with Earth, from fault-based liability, in the case of in-orbit collisions.

The Space Liability Convention has never been formally invoked. When a Soviet satellite, Kosmos 954, crashed in Canada, Canada referenced the convention but did not formally invoke it, in a claim for C$6 million for environmental damage. In the end, the U.S.S.R paid C$3 million *ex gratia*.[19]

---

[17] Secure World Foundation 'Space Sustainability: A Practical Guide' [2014] <https://swfound.org/media/121399/swf_space_sustainability-a_practical_guide_2014__1_.pdf>   accessed   24th September 2018.

[18] Joanne Gabrynowicz 'A Chronological Survey of the Development of Art. IX of the Outer Space Treaty' [2010] Supplement to Journal of Space Law, University of Mississippi School of Law <www.spacelaw.olemiss.edu/resources/pdfs/article-ix.pdf> accessed 24th September 2018.

[19] Ram Jakhu 'Iridium-Cosmos Collision and its Implications for Space Operations' in *ESPI Yearbook on Space Policy. 2008/2009: Setting New Trends* 254 (Springer Wien New York 2010).

This case highlighted the differences in interpretations of the convention, the ambiguity relating to 'damage' and the expectations of different states with regard to forewarning and reimbursement.

3.2.3 Inter-Agency Space Debris Coordination Committee

The IADC is an international governmental forum coordinating efforts to deal with space debris. Founded in 1993, the organisation currently comprises space agencies of 12 individual countries, as well as the European Space Agency (hereinafter ESA). The organisation has four working groups covering different aspects of space debris.

Space Debris Mitigation Guidelines 2007

In 1999, the IADC began developing the first international consensus on space debris mitigation, first published in 2002.

The document defines LEO and outlines mitigation measures, which organisations are encouraged to adhere to. These include (1) limitation of debris released during normal operations, (2) minimisation of the potential for on-orbit break-ups, (3) post-mission disposal and (4) prevention of on-orbit collisions. The guidelines also recommend a 25-year post-mission maximum lifetime and a 30-year maximum post-launch lifetime. Operators are encouraged to apply the guidelines 'to the greatest extent possible'.[20]

Presented to the UNCOPUOS in 2003, the IADC guidelines were used as a guideline for the *UNCOPUOS Space Debris Mitigation Guidelines* document, which was published in 2007, and subsequently endorsed by the General Assembly.

The IADC noted in 2017 that in LEO, the current implementation level of the guidelines was 'considered insufficient and no apparent trend towards a better implementation is observed'.[21]

In 2018, the UNCOPUOS General Assembly[22] noted 'with satisfaction' that some states were implementing the guideline measures. Some delegations to the assembly noted that making the mitigation guidelines legally binding was not necessary as it was in the interests of all states to preserve the sustainability of outer space.

---

[20] *IADC* (n 7) 5.

[21] IADC 'An Overview of the IADC Annual Activities' [2017] <https://www.iadc-online.org/Documents/IADC-17-02%20UN%20COPUOS%20STSC%20Presentation%20-%2021.12.2016.pdf> accessed 27th September 2018.

[22] UN GA COPOUS 'Report of the Legal Subcommittee on its fifty-seventh session' [2018] A/AC.105/1177 <https://cms.unov.org/dcpms2/api/finaldocuments?Language=en&Symbol=A/AC.105/1177> accessed 24th September 2018.

The 25-year guideline has been adopted by national agencies, including NASA and ESA, in their mission plans. NASA and the FCC require filing of an Orbital Debris Assessment Report showing adherence to the 25-year rule.[23]

Statement on Large Constellations of Satellites

Acknowledging the emerging plans for large LEO constellations in 2015, the IADC began investigating the potential associated risks. Due to the rate of the implementation of these constellations, they published the 'IADC Statement on Large Constellations of Satellites in Low Earth Orbit' in 2017 but continue to work on more comprehensive modelling.[24] In this initial analysis, the IADC says that the Space Debris Mitigation Guidelines may not go far enough, in particular that the 25-year post-mission lifetime may need to be reduced.

They note that reliability of critical systems is a 'key consideration' and 'significant improvements in the reliability of the disposal function at end of life will be needed'. Furthermore, they recommend 'Design for sufficient on-board redundancies of all functions involved in the post mission disposal'.

Regarding trackability, they recommend 'to enhance trackability by adding onboard active and/or passive components'. Whilst this is an essential step in the right direction, one could argue that the least trackable satellites, picosatellites, are the ones least likely to have additional trackability systems as the systems would take up more space, power and financial cost.

3.3   A Case Study: Iridium-Kosmos

The good news for the space community, and the bad news for space lawyers, is that there are only a handful of incidents and cases to discuss. The most relevant to this article is the Iridium-Kosmos satellite collision—the only collision between two satellites that has ever occurred.

US-based company Iridium LLC operates a LEO constellation of satellite for communications. Iridium 33 was one such satellite, launched in 1997. Kosmos 2251 was a Russian government communication satellite, launched in 1993, that had ceased functioning. The two satellites collided in 2009, creating over 2000 pieces debris and a legal question that has yet to be satisfactorily answered. Some key legal issues were raised by the incident.

[23] National Aeronautics and Space Administration 'Handbook for Limiting Orbital Debris' 2008 <https://explorers.larc.nasa.gov/APMIDEX2016/MO/pdf_files/NHBK871914.pdf> accessed 27th September 2018.

[24] Inter-Agency Space Debris Coordination Committee 'Statement on Large Constellations of Satellites in Low Earth Orbit' 2017 <https://www.iadc-online.org/Documents/IADC%20 Statement%20on%20Large%20Constellations%20rev%203.pdf> accessed 25th September 2018.

3.3.1 Launching State

Iridium 33 was not registered with the United Nations as required by the 1974 Registration Convention,[25] raising questions about who the launching state was.

The liability convention defines a state: 'A State which launches or procures the launching of a space object [or] A State from whose territory or facility a space object is launched'. The private-company-built satellite was launched by a Russian Proton launch vehicle from Kazakhstan. Iridium LLC is registered in the US, but that is the only link to the US state. It could be argued that Kazakhstan, or even Russia, was the launching state.

3.3.2 Fault

The Liability Convention requires fault to be determined for liability. *Prima facie* this seems like a logical approach; however, the lack of a definition for fault caused controversy.

Russia pointed out that Kosmos 2251 was not functioning, so it must have been Iridium's fault for not manoeuvring their satellite. Russia also noted that there was no obligation on its part to dispose of Kosmos 2251 after it stopped functioning. Iridium pointed out that there is no obligation to avoid collisions even if it was aware that such a collision would occur.[26]

Fault is defined by Black's Law Dictionary[27] as follows:

> 1. An Error or defect of judgement or conduct; any deviation from prudence or duty resulting from inattention, incapacity, perseverity, bad faith, or mismanagement ... 2. The intentional or negligent failure to maintain some standard of conduct that failure results in harm to another person.

Under this definition, it could be argued that Iridium LLC was negligent in its duty to avoid other satellites to avoid harm, or Russia was negligent in allowing its satellite to fail and descend uncontrolled into Iridium's orbit.

---

[25] Secure World Foundation '2009 Iridium-Cosmos Collision Fact Sheet' [2010] <https://swfound.org/media/6575/swf_iridium_cosmos_collision_fact_sheet_updated_2012.pdf> accessed 25th September 2018.

[26] Michael Listner 'Iridium 33 and Cosmos 2251 three years later: where are we now?' [2012] <http://www.thespacereview.com/article/2023/1> accessed 26th September 2018.

[27] *Black's Law Dictionary* (8th ed. 2004), 641.

3.3.3 Tracking

Both the US's Joint Space Operations Centre (hereinafter JSpOC) and Russia performed conjunction assessments that did not place the potential collision as high risk. The US had been providing daily warnings to Iridium, which operated over 70 satellites at the time. Due to the amount of warnings and inaccuracy in the data provided, the warnings were stopped 2 years prior to the collision.[28]

3.3.4 Lessons Learnt

No party met the standard of fault, and neither party invoked legal proceedings. The lack of claim from either side could be seen as the players hedging their bets[29]; the creation of over 2000 debris pieces increases the risk of one of these colliding with another satellite, which would, legally, cause the same questions to be raised.

After the collision, the JSpOC began daily conjunction assessments of Iridium's constellation again. JSpOC currently provide 72-h in advance warnings to operators with satellites that would conjunct within 1km of other satellites in LEO.

However, in the meantime, the lessons of Iridium Kosmos appear to have been learnt, and organisations are more rigorous with conjunction assessments. The incident also increased the public's awareness of space debris and space situational awareness. However, this accident happened with a constellation of less than 80 satellites. Constellations with orders of magnitude more satellites will require timely and accurate information from all other satellite operators in nearby orbits, and accurate data on space debris, to coordinate their constellations. Performing avoidance manoeuvres uses up fuel and power and is therefore expensive for operators.

3.4 The Future

This section will discuss developments that could be made to further legally secure the sustainability of space.

---

[28] John Campbell 'Forum on National Security Space Examining Codes and Rules for Space' [2017] <www.researchgate.net/publication/242327287_Forum_on_National_Security_Space_Examining_Codes_and_Rules_for_Space> accessed 27th September 2018.

[29] Frans von der Dunk 'Too-Close Encounters of the Third Party Kind: Will the Liability Convention Stand the Test of the Cosmos 2251-Iridium 33 Collision?' [2010] Space, Cyber, and Telecommunications Law Program Faculty Publications 28 <https://digitalcommons.unl.edu/cgi/viewcontent.cgi?article=1027&context=spacelaw> accessed 24th September 2018.

3.4.1 Space Debris Definition

The definition of space debris must be established beyond doubt—this could be through international treaty or international customary law. This would allow the Liability Convention to be more useful—fault could be established by the definition and the mitigation efforts undertaken by the launching state.

The UNCOPUOS General Assembly[30] has noted that there should be a consultative process on the definition of space debris.

3.4.2 Fault Definition

The Liability Convention has no formal definition of fault; this is problematic. At what point is insufficient design a fault? Or failure to have sufficient active tracking mechanisms? Is non-compliance with guidelines considered fault? The lack of clarity highlighted by Iridium-Kosmos must be rectified before a similar event occurs.

3.4.3 Orbit Selection Management

Currently this has been workable—combined with the Liability Convention, states have taken responsibility for their own spacecraft not conflicting with others despite there being no formal organisation that designates orbits in the way the International Telecommunications Union designates spectrum. However, if specific orbits are sought after or deemed crowded, regulation in the same way as geostationary orbits could be considered.

A centralised system for orbit selection would help in future liability cases. If constellation systems are assigned specific separate orbits, it could be shown that system A had strayed into system B's orbit and therefore system A is at fault. This would work regardless of whether the space object was an active object or debris, though further clarification of whether debris is liable for collisions, highlighted by Russia in Iridium-Kosmos, is necessary.

---

[30] UNCOPOUS General Assembly A/AC.105/1177 [2018].

3.4.4   Active Debris Removal

As technologies develop that look to remove objects from space, this should be considered from a legal perspective. Only about 20% of satellites above 650 km attempt to deorbit,[31] so many will require an active debris removal method if they are to return to Earth. Currently, a removing actor would have to seek permission from the state that owns the space debris. This could be an unnecessary hurdle to clean up space.

One could look to other international legal domains to find solutions for these problems—for example, the maritime law contains the concept of salvage.

The law of salvage is a maritime law principle that entitles a person who recovers another person's property at sea to be rewarded commensurate with the value of the property. Currently this is not directly analogous to space because recovered satellites are unlikely to have much value beyond being museum artefacts. However, if space debris were legally defined, salvors could claim to be acting in the interests of the international community in removing it and need not have to seek permission from the owner. This would also be beneficial for instances where the space debris is unidentified or ownership is contested.

3.4.5   Tracking of Objects

Whilst the registration convention ensures that space objects are registered at an international level, there is no system for the tracking of space objects. Currently this is done by individual governments. This poses a question regarding the reliability of ownership of objects, including debris. It is difficult to know for certain who owns objects in space that could have been launched half a century ago. New companies are developing systems for more accurate tracking; however, there will always be an issue of accountability of the data.

3.4.6   Space Debris Mitigation Guidelines Becoming Customary International Law

Customary international law is a source of international law applicable to all states.[32] If an obligation is sufficiently general and widespread and evidence of *opinio juris* across states to which the practice applies, an international law obligation can

---

[31] Jeff Foust 'Mega-Constellations and Mega- Debris' [2016] The Space Review <http://www.thespacereview.com/article/3078/1> accessed 25th September 2018.

[32] United Nations, 'Charter of the United Nations and Statute of the International Court of Justice' [1945] Article 38(1)(b).

become customary international law.[33] The adoption of the Space Debris Mitigation Guidelines could be argued to be customary international law, despite their infancy.[34] This would mean that the lack of specific treaty pertaining to these issues is not a problem, as a claimant could turn to existing practices across nations as evidence of *opinio juris*.

3.4.7 Ergo omnes *Obligations*

The growth of the 'NewSpace' industry has been due to the increase in accessibility of space. Whilst it is unlikely at an international level, over-regulation would be damaging to the sector. Navigating, through monitoring and avoidance manoeuvres, space debris is expensive—more expensive than adhering to guidelines. It is in the interests of all space actors that space debris be avoided, especially in constellation orbits. Actors have *ergo omnes* obligations to adhere to current and future recommendations. Clarified regulations would be of benefit to both governments and companies. Guidelines, such as the Space Debris Mitigation Guidelines, though initial adoption is low, are recommended for this.

3.5 Conclusion

This article has outlined the relevant legal aspects of space debris pertaining to satellite constellations and identified areas for clarification in the existing legal framework.

It is easy to pose hypothetical questions and far more difficult to answer them. Space law requires the marrying of complex technical innovations, delicate international relations and myriad national legislation. In the development of space law, the clock turns slowly, but the emergence of new players in the space industry and fast iterations of constellation design and launch require new answers faster than ever before. Space has been a pioneer for international cooperation, with joint projects such as the International Space Station demonstrating what can be done when international cooperation is underpinned by international law. One hopes that this can be continued into the future through the continued sustainability of this shared resource.

Ewan Wright is currently studying a master's degree in Aerospace Engineering at the University of Sheffield. As part of his degree, he is currently undertaking a Year in Industry placement at the Satellite Applications Catapult in Harwell, UK. He has a particular interest in small satellites and future constellations. Contact via email: etwright1@sheffield.ac.uk.

---

[33] *Nicaragua v. United States of America* International Court of Justice [1986] <www.icj-cij.org/files/case-related/70/070-19860627-JUD-01-00-EN.pdf> accessed 26th September 2018.

[34] Scott Kerr 'Liability for Space Debris Collisions and the Kessler Syndrome' [2017] <http://www.thespacereview.com/article/3387/1> accessed 26th September 2018.

Chapter 4
The Rise of the LEO: Is There a Need to Create a Distinct Legal Regime for Constellations of Satellites?

Alice Rivière

Abstract Recently, various entrepreneurs have revived the idea of constellations of hundreds and even thousands of satellites orbiting in the low-earth orbit (LEO), which would lead to a dramatic increase in the number of satellites deployed in this orbit. The main issue arising out of those massive projects is the congestion of the LEO and its legal consequence: the question of spectrum management and regulation and radio frequency allocation in the LEO and the potential need to adapt a legal frame for the specificity of the LEO.

By analyzing the elements that led to the creation of a distinct legal regime for the geostationary orbit (GEO), we shall determine whether there is a need to create such a distinct legal regime for the LEO.

4.1 A Nascent Technological Figure in the Satellite Realm

By the turn of this century the new LEO technology will rise and soar in the outer space over every country on the earth. Dozens of times each day fleets of LEO satellites will sail across the sky, blasting messages everywhere, ferrying countless communications, and delivering endless information, poetically predicted in 1995 by famous American lawyer and entrepreneur Martine Rothblatt.[1]

The views and opinions expressed in this essay are the ones of its author and do not reflect the views of Airbus Defence and Space.

[1] Martine Rothblatt, "Lex Americana: The New International Legal Regime For Low Earth Orbit Satellite Communications Systems", Journal of Space Law, Vol. 23, No. 2 (1995), p. 140.

A. Rivière (✉)
Airbus Defence and Space in Munich, Taufkirchen, Germany

© Springer Nature Switzerland AG 2019
A. Froehlich (ed.), *Legal Aspects Around Satellite Constellations*, Studies in Space Policy 19, https://doi.org/10.1007/978-3-030-06028-2_4

In the 1990s, a new technological figure emerged in the world of satellites: the constellation of computerized satellites deployed in strategic orbital positions to form an interconnected satellite network able to provide global personal communication systems.[2] Satellite constellations present two particularities: first, as their name implies, they are deployed in an interconnected nongeostationary fleet, meaning that the position of the satellites changes in relation to the surface of the earth to form a moving constellation, circling the globe and relaying messages back and forth between each other and users and/or earth stations.[3] Second, they are deployed at a low orbital altitude.[4]

Habitually, communication satellites were launched at a much higher orbital altitude, in the so-called geostationary orbit (GEO), *i.e.* a circular orbit located at approximately 36,000 km above the equator.[5] The peculiarity of satellites deployed in this orbital slot, which circular and direct orbit lie in the place of the earth's equator, is that they revolve around the earth at the same speed as the rotation of the earth.[6] When observed from a given point on the surface of the earth, it gives the impression that they remain stationary.[7] In reality, they move in a figure-eight pattern within the orbit volume.[8] This effect is important because those seemingly stationary satellites can be used as a communication repeater between two points on the surface of the earth without having to move the emitting or receiving earth station or without requiring the antenna at those stations to follow the movements of this satellite.[9] This is the reason why mobile satellite communications were initially deployed in the GEO.[10]

---

[2] Martine Rothblatt, Lex Americana: The New International Legal Regime For Low Earth Orbit Satellite Communications Systems, Journal of Space Law, Vol. 23, No. 2 (1995), p. 140.

[3] ITU Press Archiv WRC97, What Are the New Mobile Satellite Systems?, available at: https://www.itu.int/newsarchive/press/WRC97/What-are-new-mss.html (accessed on September 30, 2018).

[4] Martine Rothblatt, Lex Americana: The New International Legal Regime For Low Earth Orbit Satellite Communications Systems, Journal of Space Law, Vol. 23, No. 2 (1995), p. 125.

[5] See Article 1.188 ITU Radio Regulations and C. Brandon Halstead, Promethues Unbound – Proposal for a New Legal Paradigm for Air Law and Space Law: Orbit Law, Journal of Space Law, Vol. 36, 2010, p. 172.

[6] S. Hobe, Geostationary Orbit, in: Max Planck Encyclopedia of Public International Law, 2007, para. 2: "Operating in the GEO, satellites orbit the earth in 23 hours, 56 minutes, and 4.2 seconds, revolving at a rate that is synchronous to the rotation of the earth."

[7] R. Jakhu, The Legal Status of the Geostationary Orbit, Annals of Air and Space Law, Vol. 7, 1982, p. 333.

[8] S. Hobe, Geostationary Orbit, in: Max Planck Encyclopedia of Public International Law, 2007, para. 3.

[9] E. DuCharme, R. Bowen, M. Irwin, The Genesis of the 1985/87 ITU World Administrative Radio Conference on the Use of the Geostationary-Satellite Orbit and the Planning (1) of Spaces Services Utilizing It, Annals of Air and Space Law, Vol. 7, 1982, p. 262.

[10] The first MARISAT mobile communications satellite was launched into a GEO over the Pacific Ocean in 1976 to provide communications between ships and shore stations. See G. Comparetto, N, Hulkower, Global Mobile Satellite Communications: a review of three contenders, originally presented at the 15th International Communication Satellite Systems Conference of San Diego, 1994, p. 2.

However, at the dawn of the 1990s, when mobile communication systems were emerging, significant technological progresses for space-based satellite-based mobile communication systems were made, such as digital voice processing, satellite technology, and component miniaturization.[11] These innovations allowed some companies to develop mature proposals for such systems based on constellations of satellites to be deployed into the LEO. Orbit where the satellites remain lower than 2000 km above the earth's surface.[12] The idea was to launch those constellations at lower orbital altitude to improve signal quality, reduce the time delay of transmission, and require less booster power.[13] Because the lower altitude of satellites in the LEO shrinks their respective footprint, it is necessary to deploy them within a constellation in several orbital plane to achieve global coverage.[14]

However, those ambitious and groundbreaking constellation projects did not live up to their promises, and toward the end of the 1990s, most of the projects had failed and had been abandoned due to financial reasons.[15]

Those failures prompted commentators to conclude that LEO satellite constellations were "conceived in the age of voice and woefully underequipped in today's age of data."[16] Yet almost 20 years later, various entrepreneurs have revived the idea of constellations of hundreds and even thousands of satellites based on new technologies that now enable the fabrication of smaller, lighter, less expensive, but still very capable satellites.[17] Recently, five companies have announced their ambitious plans to deploy thousands of Internet-service nongeostationary satellites in

---

[11] G. Comparetto, N, Hulkower, Global Mobile Satellite Communications: a review of three contenders, originally presented at the 15th International Communication Satellite Systems Conference of San Diego, 1994, p. 1.

[12] The proposal for a LEO satellites constellation where initiated by several companies. In 1998, Orbcomm became the first commercial provider of global LEO satellite date and messaging communications services, after deploying a constellation of 28 satellites in LEO. The Motorola Corporation planned a constellation named "Iridium" containing 66 satellites to establish a satellite based phone system that was eventually deployed in 1998. The Teledesic Corporation initially developed a plan for a global satellite communications network using 840 small satellites with in-orbit spares at an altitude of 700 km which was scaled back in 1997 to 288 active satellites at 1400 km. Globalstar had a similar idea in nature and deployed a constellation of 48 satellites at 1414 km in the LEO between 1998 and 2000. SkyBridge, Alcatel's satellite-based broadband access system, planned a constellation of 64 LEO satellites at an altitude of 1457 km.

[13] P. Sourisse, D. Rouffet and H. Sorre, SkyBridge: a broadband access system using a constellation of LEO satellites, ITU Press Archiv WRC97, available at: https://www.itu.int/newsarchive/press/WRC97/SkyBridge.html (accessed on September 30, 2018).

[14] See LEO Vantage 1 (Telesat LEO Phase 1) Satellite Presentation, available at: http://spaceflight101.com/pslv-c40/leo-vantage-1/ (link accessed on September 30, 2018).

[15] By the beginning of 2000, OrbComm, Iridium, and Globalstar had all gone bankrupt and filed for Chapter 11 protection in the U.S. while Teledesic suspended its activities in 2002, see R. Goodwins, Teledesic backs away from satellite push, 2002, available at: https://www.zdnet.com/article/teledesic-backs-away-from-satellite-push/ (link accessed on September 30, 2018).

[16] C. Melow, The Rise and Fall and Rise of Iridium, Air & Space Magazine, September 2014, p. 3.

[17] W. Ailor, G. Peterson, J. Womack, M. Youngs, Effect of large constellations on lifetime of satellites in low earth orbits, p. 119.

constellation in LEO.[18] These constellations are specifically designed to deliver high-speed Internet access to urban, suburban, and rural areas which are not yet connected to broadband terrestrial infrastructure or which are uneconomical to cover with traditional infrastructure.[19] Indeed, LEO satellite fleets appear very promising for business models and services where the decisive element is the price of the connection, as is the case for global Internet coverage or for the Internet of Things.[20] Moreover, existing constellations launched in the nineties are being replaced with second-generation satellites.[21]

Accordingly, it appears that the LEO has reemerged as the new paradigm for satellite communications and that global broadband Internet delivered via nongeo-stationary satellite orbit is the next emerging capability within the commercial space sector.[22]

4.2 The High Number of Launches Planned Raises Issue of Spectrum Management and Regulation for the LEO

4.2.1 A Dramatic Increase in the LEO Satellite Population

Given the amount of expected space launches planned in the near future, one can wonder about the sustainability of the orbital environment at this altitude for space operations and consequently question the launch and deployment of such a high number of satellites, which could result in the congestion or saturation of the LEO. Indeed, should these constellations be realized, the number of operating

---

[18] Elon Musk's SpaceX was the first US-based entity authorized by the American Federal Communications Commission ("FCC") to launch and operate Starlink, a massive broadband internet satellite constellation of 4425 broadband satellites orbiting approximately between 1100 and 1325 km in LEO; its concurrent OneWeb petitioned the FCC to access the US market with thousands of satellites authorized by the UK; Canadian satellite communication service provider Telesat also has a satellite internet project and plans deploy a constellation of 117 satellites in the LEO which was approved by the FCC in November 2017; Boeing and Leosat have also announced ambitious plans to put thousands of Internet-service satellites in the LEO.

[19] See e.g. OneWeb's plan to cover remote regions such as Alaska and other Artic region as emphasized in a letter from Bill walker, Governor, State of Alaska, to the Honorable Tom Wheeler, Chairman, FCC, filed August 3. 2016: "I am particularly enthused about the fact that OneWeb's network of satellites will be in a pole to pole direction, such that Alaska and other Artic regions will receive tremendous coverage."

[20] B. Trevdic, Satellites: la bataille des orbites divise le secteur, Les Echos, March 19, 2015, p. 1.

[21] See e.g. Iridium CEO Matt Desch's announcement on October 18, 2018 that the last mission needed to complete Iridium's second-generation satellite constellation was scheduled for December 30, 2018.

[22] S. Le May, S. Gehly, B. Carter, S. Flegel, Space debris collision probability analysis for proposed global broadband constellations, Acta Astronautica, Vol. 151, 2018, p. 445.

satellites in the LEO regime would dramatically increase—from ~1200 now to over 6000.[23]

With so many satellites orbiting in different planes and at different altitudes, the main issue arising out of those massive projects is the issue of congestion of the LEO, which has two legal consequences: the question of the management and regulation of spectrum and radio frequency allocation in the LEO and the need to adapt the legal frame for the specificity of the LEO and its pendant, the issue of space debris. This essay will concentrate only on the first aspect.

4.2.2   Risks of Congestion and Saturation

The radio-frequency spectrum is a finite natural resource, and there is a risk of shortage of spectrum because, although radiation is practically unlimited, the information a spectrum can transport depends on technological capability.[24] In addition, positions available in the LEO are limited.[25]

Potentially, the deployment of such a high number of satellites raises issues of spectrum shortage, interferences with competing systems up to the point of possible collision between satellites but also with existing systems in the GEO and, lastly, with radio astronomy services.[26]

In the realm of satellites, risks of interference are named crosstalk or inline events and occur when two satellites are simultaneously above an area that both are communicating with so that their beams cross, causing potential radio interference.[27]

[23] W. Ailor, G. Peterson, J. Womack, M. Youngs, Effect of large constellations on lifetime of satellites in low earth orbits, p. 119. In addition to the tracked objects, there are estimates that as many as 8000–10,000 additional debris objects larger than 10 cm also reside in LEOs as well as an estimated 500,000 objects in the range 1–10 cm in size and millions of still smaller objects, *see* W. Ailor, G. Peterson, J. Womack, M. Youngs, Effect of large constellations on lifetime of satellites in low earth orbits, p. 117.

[24] ITU News Archive WRC97; What are the New Mobile Satellite Systems, available at: https://www.itu.int/newsarchive/press/WRC97/What-are-new-mss.html (accessed on September 30; 2018).

[25] P. Arnopoulos, The International Politics of the Orbit-Spectrum Issue, Annals of Air and Space Law, Vol. 7, 1982, p. 217.

[26] The U.S. National Radio Astronomy Observatory expressed concerns about possible interference to radio astronomy facilities before the FCC in the context of the licence to access the U.S. market applications of the satellites constellations.

[27] See e.g. "An "in-line" event occurs when satellites of different NGSO FSS systems are physically aligned with an operating earth station of one of those systems, such that the topocentric angle between the satellites is less than 10 degrees as measured from the earth station.", FCC, Notice of Proposed Rulemaking, December 15, 2016, para. 22.

In the scientific literature, the risk of collision in an overcrowded LEO, also known as the "Kessler Syndrome," was envisaged very early on.[28] As early as 1978, scientists started worrying that the LEO would one day become unusable, positing that at a certain point, a cascade of collisions will envelop the earth and close off access to certain areas of space, including the LEO.[29]

Admittedly, as of today, there have been at least four reported accidental hyper-velocity collision events in the LEO, including a collision between satellites of the first generation's LEO constellations: the collision of Iridium 33 and defunct Russian satellite Cosmos 2251.[30] A study even concluded that there is a high probability for the occurrence of at least one collision for both the proposed OneWeb and SpaceX constellations during an operational phase of 5 years.[31]

4.3 Comparison with the GEO

4.3.1 An Early Awareness for the Scarcity, Limitedness, and Saturation of the GEO

Concerns of saturation and congestion are not unique to the LEO; when satellite deployment in the GEO multiplied and accelerated, similar solicitude over the congestion of the GEO was raised. The need to optimize the development and equitable sharing of the GEO and address the problems of its limitedness, scarcity, and saturation was recognized from the outset by the International Telecommunication Union (ITU), the United Nations specialized agency for information and communication technologies and international regulator of the satellite industry,[32] devoted to the technical facilitation of international telecommunication.[33] Originally, the ITU regulated terrestrial radio services but the principles it developed in this context were very early on transposed to the regulation of space activities, and in 1973 its powers were enlarged to include the management of orbital positions and the use of the

---

[28] D. J. Kessler, Collisional cascading: the limits of population growth in low Earth orbit, Advance Space Research, Volume II, 1991, pp. 3–6.

[29] D. St. John, The Trouble with Westphalia in Space: The State-Centric Liability Regime, Denver Journal of International Law and Policy, Vol. 40, 2012, p. 688.

[30] At 16:56 UTC on February 10, 2009, *Iridium 33* collided with the defunct Russian satellite Kosmos 2251, *see* S. Le May, S. Gehly, B. Carter, S. Flegel, Space debris collision probability analysis for proposed global broadband constellations, Acta Astronautica, Vol. 151, 2018, p. 445.

[31] S. Le May, S. Gehly, B: Carter, S. Flegel, Space debris collision probability analysis for proposed global broadband constellations, Acta Astronautica 151 (2018), 453.

[32] R. Jakhu, The Legal Status of the Geostationary Orbit, Annals of Air and Space Law, Vol. 7, 1982, p. 344.

[33] F. Lyall, The International Telecommunication Union and Development, Journal of Space Law, Vol. 22, 1994, p. 24.

radio-frequency spectrum.[34] Notably, the ITU regularly organizes World Radiocommunication Conferences ("WRC," known until 1992 as "WARC," World Administrative Radio Conferences), which are intended to examine the need for frequency allocation and attempt to apportion the spectrum in an equitable and forward-looking way while protecting the services already in place.[35]

The first geostationary satellite was put into orbit in 1963. The same year, the ITU held an Extraordinary Administrative Radio Conference (EARC) to develop the basic administrative and technical regulations for the operation of space systems and allocate frequencies to the various space radiocommunication services. From the outset, the debate was politically tainted and the aim was to ensure that "communication satellites should be organized on a global basis with nondiscriminatory access for all nations."[36]

4.3.2 A Progressive Recognition of the GEO as a Limited Natural Resource Despite a Heated Political Debate

Progressively, the distinctive nature of the GEO as a unique part of outer space deserving a special legal status was recognized.[37] This is reflected in the consensus reached in Article 33(2) of the International Telecommunications Convention of 1973 stating that the GEO is a limited natural resource that must be used economically and efficiently to allow equitable access to it by all countries.[38]

Over the years, the debate around the scarcity of available GEO positions turned into a heated international debate around the question distribution among states against the background of a North/South polarization.[39] This polarization culmi-

---

[34] P. De Man, Rights over Areas vs Resources in outer Space: What's the Use of Orbital Slots, Journal of Outer Space, Vol. 38, 2012, p. 43.

[35] ITU News Archive WRC97, What are the New Mobile Satellite Systems, available at: https://www.itu.int/newsarchive/press/WRC97/What-are-new-mss.html (accessed on September 30; 2018).

[36] *See* U.N. General Assembly Resolutions 1721 (XVI) part D and 1802 (XVII) part IV; *See also* E. DuCharme, R. Bowen, M. Irwin, The Genesis of the 1985/87 ITU World Administrative Radio Conference on the Use of the Geostationary-Satellite Orbit and the Planning (1) of Spaces Services Utilizing It, Annals of Air and Space Law, Vol. 7, 1982, p. 265. It was at that occasion that the backbone principle of the ITU registration, i.e. the priority principle, oftentimes labelled the principle of first come, first served, was first introduced.

[37] R. Jakhu, The Legal Status of the Geostationary Orbit, Annals of Air and Space Law, Vol. 7, 1982, p. 349.

[38] Article 33(2) of the ITU Convention provides that: "In using frequency bands for the space radio services Members shall bear in mind that radio frequencies and the geostationary orbit are limited natural resources, that they must be used efficiently and economically so that countries or group of countries may have equitable access to both in conformity with the provisions of the Radio Regulations according to their needs and the technical facilities at their disposal."

[39] P. Arnopoulos, The International Politics of the Orbit-Spectrum Issue, Annals of Air and Space Law, Vol. 7, 1982, pp. 219 and 232.

nated with the Bogota Declaration of 1976 when a group of equatorial countries denied the generally held view that the GEO is located in outer space and that the application of Article II of the Outer Space Treaty of 1967, which stipulates that outer space, unlike air space which is under national sovereignty, is not subject to national appropriation by claim or by occupation.[40] Even if this submission about the GEO's location was vehemently rejected as a whole by the international community,[41] the need to establish a special legal regime to govern the use of this orbit was widely recognized.[42]

4.3.3 Creation of a Specific Legal Regime for the GEO

Over the years, administrative regulations regarding spectrum sharing for GEO satellites were developed and sophisticated coordination and notification procedures were introduced.[43] However, coordination between governmental users of the GEO eventually proved to be difficult to achieve: when India and Indonesia were planning their respective domestic satellite systems in the GEO, other nations in that orbit did not readily agree to adjust their operational systems.[44] India and Indonesia then sought to have this issue dealt with before the ITU.[45] A special WARC, held in two sessions in 1985 and 1987, was thus convened to specifically address the GEO. The outcome of the conference resulted in the allotment to each state, whether or not a member of the ITU, of a nominal orbital position within a predetermined arc and a predetermined band of the GEO.[46] To secure the position of nonspacefaring nations, it was decided that even if a state that initially decided not to use its

---

[40] R. Jakhu, The Legal Status of the Geostationary Orbit, Annals of Air and Space Law, Vol. 7, 1982, p. 335.

[41] See B. Brittingham, Does the World Really Need New Space Law, Oregon Review of International Law, Vol. 12, 2010, p. 46. In particular, the Bogota Declaration was not discussed during any of the next ITU conferences, the ITU considering that the question concerned the UN Committee on the Peaceful Uses of Outer Space.

[42] Such as Canada, Sweden, Spain, and the Netherlands. See R. Jakhu, The Legal Status of the Geostationary Orbit, Annals of Air and Space Law, Vol. 7, 1982, p. 342.

[43] See Article 9A of the Radio Regulations implementing a procedure of advance publication of information on planned geostationary satellite systems via the ITU's International Frequency Registration Board.

[44] S. Doyle, Space Law and the Geostationary orbit : the ITU's Warc-Orb 85-88 Concluded, Journal of Space Law, Vol. 17, 1989, p. 13.

[45] Doyle, Space Law and the Geostationary orbit: the ITU's Warc-Orb 85-88 Concluded, Journal of Space Law, Vol. 17, 1989, p. 14.

[46] S. Doyle, Space Law and the Geostationary Orbit: the ITU's Warc-Orb 85-88 Concluded, Journal of Space Law, Vol. 17, 1989, p. 15. Those rules have later been integrated into the ITU Convention and Constitution from December 22, 1992.

position on the GEO arc suddenly changed its mind, it still had a right to take it from prior occupants.[47]

Finally, the unique status of the GEO has been enshrined in Article 44 of the ITU Constitution of 1992, which calls upon member states to acknowledge that the radio-frequency spectrum and satellite orbits, including the geostationary-satellite orbit, are limited natural resources that "must be used rationally, efficiently and economically, in conformity with the provisions of the Radio Regulations, so that countries or group of countries may have equitable access to those orbits and frequencies, taking into account the special needs of the developing countries and the geographical situation of particular countries."[48] The Preamble of the Radio Regulations establishes similar principles."[49] The phrasing of Article 44(2) has been interpreted as implying that "the efficient and economic use of orbits is a prerequisite condition for attaining the ultimate yet necessarily subsequent goal of equitable access."[50]

4.4   Developments That Occured for the GEO Are Not Transposable to the LEO

4.4.1   A Different Political Context

When comparing those evolutions concerning the GEO with what might be the case with the LEO, one aspect is very striking: the political context of the emergence of LEO satellites is radically different from the political context of New World Order that surrounded the emergence of GEO satellites. Foremost, new state players have emerged in the space industries: Asia has joined Europe, Russia, and North America

---

[47] F. Lyall, The International Telecommunication Union and Development, Journal of Space Law, Vol. 22, p. 30.

[48] Article 44 of the ITU Constitution (Use of the Radio-Frequency Spectrum and of the Geostationary-Satellite and Other Satellite Orbits) reads as follows:

1. Member States shall endeavour to limit the number of frequencies and the spectrum used to the minimum essential to provide in a satisfactory manner the necessary services. To that end, they shall endeavour to apply the latest technical advances as soon as possible.
2. In using frequency bands for radio services, Member States shall bear in mind that radio frequencies and any associated orbits, including the geostationary-satellite orbit, are limited natural resources and that they must be used rationally, efficiently and economically, in conformity with the provisions of the Radio Regulations, so that countries or groups of countries may have equitable access to those orbits and frequencies, taking into account the special needs of the developing countries and the geographical situation of particular countries.

[49] J. Zoller, Improving the International Satellite Regulatory Framework.

[50] P. De Man, Rights over Areas vs Resources in outer Space: What's the Use of Orbital Slots, Journal of Space Law, Vol. 38, 2012, p. 47.

as active participants in space, and new spacefaring nations such as China, India, and Japan have emerged.[51] This is evidenced by the latest update record of the Union of Concerned Scientists, which keeps a record of the operational satellites: as of April 2018, nearly one half of the 1186 LEO satellites are associated with the U.S, while China and Russia both have launched a considerable amount of LEO satellites and many other countries own their own LEO satellites, including Algeria, Belarus, Indonesia, Iran, Malaysia, Pakistan, Peru, Nigeria, Vietnam, Venezuela, Ukraine, United Arab Emirates, Turkey, and Sweden.[52]

4.4.2   Key Players Are Private Actors Aiming at Providing Global Coverage

Another aspect is particularly noteworthy: while at the time of the development of GEO satellites states were the primary space actors,[53] private actors have now massively joined the space race and emerged as key players in space.[54] This is particularly true for LEO satellites ventures: as a matter of fact, all constellation initiatives planned and launched since the 1990s have been private initiatives. The increasingly dynamic private sector involvement induced a shift toward a polycentric approach to atmospheric management.[55] Most importantly, one of the consequences of this shift is the eradication of national sovereignty claims concerning orbital positions in the LEO. It is true that when the first LEO constellations were authorized in the U.S. by the Federal Communication Commission (FCC),[56] the European Community at that time expressed concerns about the establishment by the U.S. of a regulatory regime for LEO satellite constellations without the full involvement and

---

[51] B. Brittingham, Does the World Really Need New Space Law, Oregon Review of International Law, Vol. 12, 2010, p. 31.

[52] UCS' data based updated on April 30, 2018, available at https://www.ucsusa.org/nuclear-weapons/space-weapons/satellite-database#.W7AFTi_pM1I (accessed on September 30, 2018).

[53] It is considered that Space Services Inc. changed the *status quo* of the state-centric international regime applicable for space law in 1989 when it became the first private company to launch a satellite into orbit, see D. St. John, The Trouble with Westphalia in Space: The State-Centric Liability Regime, Denver Journal of International Law and Policy, Vol. 40, 2012, p. 687.

[54] S. Shackelford, Governing the final frontier: A polycentric approach to managing space weaponization and debris, American Business Law Journal, Vol. 51, Issue 2, 2014, p. 432.

[55] S. Shackelford, Governing the final frontier: A polycentric approach to managing space weaponization and debris, American Business Law Journal, Vol. 51, Issue 2, 2014, p. 433.

[56] Under international space law system, any outer space activity performed by a non-governmental entity must occur pursuant to the jurisdiction and control of a State. In the U.S., because federal law provides that no one may construct radio transmitting equipment except pursuant to a construction permit granted by the Federal Communication Commission ("FCC") and because LEO satellites are specifically considered radio transmitting equipment, such national authorization comes from the FCC: *See* Communications Act of 1934, as amended, 47 U.S.C. 151 and Martine Rothblatt, Lex Americana: The New International Legal Regime For Low Earth Orbit Satellite Communications Systems, Journal of Space Law, Vol. 23, No. 2 (1995), p. 140.

participation of other countries.[57] Concerns that the U.S. might dominate the LEO arena crystallized ahead of the 1995 World Radio Conference when Teledesic, one of the satellite constellation proposals of the first generation, managed to add to the agenda of the conference a proposal concerning spectrum allocation issues in the LEO, without having followed the ordinary proposal procedure.[58] Teledesic's business model helped the corporation counteract any negative reactions: when European countries started to criticize Teledesic's behavior, the latter sought support from the developing, unwired states which would be covered by its global communications network.[59] This global coverage feature is still today a very important element to recognize when comparing the issues that may arise with respect to LEO constellations with what happened in the context of the GEO. Indeed, this is an important element that differentiates what is at stake in the LEO from what is at stake in the GEO. As a matter of fact, even if most proposals for second-generation LEO constellations today come from the U.S., they all aim at rendering global Internet coverage possible and hence work for the common good and benefit of all mankind. Again, an important consequence is that it does not create room for the emergence of national sovereignty or discrimination among nations' claims concerning the LEO.

4.5   How to Best Address the Spectrum Management Issue for the LEO?

4.5.1   Development of the Legal Framework Applicable to LEO Satellites

Even if political debates concerning the sharing of LEO positions are not likely to arise at international level, it does not eliminate the issue of its limitedness, scarcity, and saturation and the general need to manage the LEO optimally, which has been recognized from the outset. In the context of the first generation of LEO constellation, the FCC determined in 1994 that there was not enough worldwide spectrum available in the LEO frequency bands for all of the applicants before it.[60]

---

[57] *See* In re Amendment of the Commission's Rules to Establish Rules and Policies Pertaining to a Mobile Satellite Service in the 1610-1626.5/2483.5-2500 MHz Frequency Bands, 9 F.C.C. Rcd. 6020-21 (1994).

[58] K. Coale, Teledesic Mounts Lead in New Space Race, Wired Magazine, 1997, available at: https://www.wired.com/1997/10/teledesic-mounts-lead-in-new-space-race/ (link accessed on September 30, 2018).

[59] K. Coale, Teledesic Mounts Lead in New Space Race, Wired Magazine, 1997, available at: https://www.wired.com/1997/10/teledesic-mounts-lead-in-new-space-race/ (link accessed on September 30, 2018).

[60] In re Amendment of the Commission's Rules to establish Rules and Policies Pertaining to a Mobile Satellite Service in the 1610-1626.5/2483.5-2500 MHz Frequency Bands, FCC Rcd. 5936, at 5954–5975.

To answer the question whether there is for the LEO, as was the case for the GEO, a need to establish and develop a separate legal regime, we must first analyze the applicable legal frame. The limited natural resource principles developed in 1973 for the GEO have been extended to any orbit, including the LEO, and the jurisdiction of the ITU over all orbits was formalized during the 1998 Minneapolis ITU Plenipotentiary Conference.[61] In addition, the obligations of Article 44 of the ITU Constitution were extended in 1992 to apply to all orbits.[62] Thus, the lawful usage of any orbital location in the LEO is also delimited by the criteria of efficiency, economy, and equity.[63] In addition, as is the case for GEO satellites, the principle of priority also applies to the LEO and the user that first notified the ITU of the use of a specific frequency and orbital position will be legally protected from interference.

Throughout the years, several WARC/WRC conferences have, little by little, addressed the regulation of LEO satellite systems and develop new rules. During the WARC held in 1992, spectrum for LEO mobile satellite service was allocated in certain frequency bands and a detailed ITU notifying protocol was adopted for all relevant technical parameters of LEO systems and for coordinating technical interference potential among such systems and other occupants of the frequency bands.[64] During the WRC held in 1997, the ITU further opened nearly 7 GHz of spectrum globally to connect the world through nongeostationary satellite systems.[65]

Notably, the most important regulatory step concerning LEO satellites was the introduction of Article 22 of the ITU Radio Regulations, which happened during the 1997 WRC and was subsequently adjusted at the 2000 WRC.[66] This provision sets hard equivalent power-flux density (EPFD) limits to protect geostationary satellites from nongeostationary and enables nongeostationary systems to share frequencies with and protect geostationary systems without requiring individual coordination with all the systems worldwide.[67] Equivalent power-flux density is an interference indicator that takes into account the aggregate of the emissions from all nongeosta-

---

[61] P. De Man, Right over Areas vs Resources in outer Space: What's the Use of Orbital Slots, Journal of Outer Space, Vol. 38, 2012, p. 44.

[62] P. De Man, Rights over Areas vs Resources in outer Space: What's the Use of Orbital Slots, Journal of Space Law, Vol. 38, 2012, p. 45.

[63] P. De Man, Rights over Areas vs Resources in outer Space: What's the Use of Orbital Slots, Journal of Space Law, Vol. 38, 2012, p. 48.

[64] M. Williams, Little LEO spectrum allocations: the Final Analysis case, WRC/97, ITU News Archive, available at: https://www.itu.int/newsarchive/press/WRC97/little-leo.html (link lastly check on September 30, 2018).

[65] S. Courteix, Droit de l'espace, in: Répertoire de droit international Dalloz, 2017, para. 38.

[66] See WRC-00, Resolution 46: Protection of geostationary fixed-satellite service and geostationary broadcasting-satellite service networks from the maximum aggregate equivalent power flux-density produced by multiple non-geostationary fixed- satellite service systems in frequency bands where equivalent power flux-density limits have been adopted.

[67] T. Kadyrov, World Radiocommunication Seminar 2016 Equivalent power flux density limits (EPFD), available at: https://www.itu.int/en/ITU-R/space/WRS16space/WRS-16_EPFD.pdf (link accessed on September 30, 2018).

tionary satellites in the direction of any geostationary earth station, taking into account the geostationary antenna activity.[68] This rule is a caveat to the ITU's priority regime for LEO satellites and does not apply to GEO services that are not required to reciprocate to protect existing LEO systems.

4.5.2 No Urge Observed to Specifically Adapt the ITU System to LEO Satellites

Thus, the only notable evolution in the international legal framework applicable to LEO satellites was the introduction of a provision designed to protect GEO satellites from LEO satellites. No urge to adapt the ITU system to new satellite fleets has been observed on the side of the ITU. Contrary to what happened in the 1970s with respect to the GEO, there has been no calls for the creation of a unique and distinct legal regime for the LEO at this stage. Even when the ITU specifically addressed the question of the new LEO constellation proposals, it did not give the impression that a whole new regulation system would be needed. Rather, it insisted on the need for "a new coordination process"[69] and spoke in favor of coordination between operators, stressing that it did not intend "to state an order of priorities for rights to a particular orbital position and the coordination process is a two way process."[70]

4.5.3 Call for Coordination to Solve Spectrum Management Issues in the LEO

Coordination has always played an essential role in the international legal space regime and is becoming increasingly critical as states and commercial ventures undertake more space missions and launch more satellites and space stations.[71] It is also a decisive feature of the ITU system which relied upon its member states to exercise goodwill and mutual assistance from the outset.[72] Additionally, considering that the triggering event for the development of a separate legal regime for the GEO was the impossibility to achieve cooperation, the current position of the ITU appears logical. Besides, operators expressed a great eagerness to cooperate in relation with their LEO constellation proposals. SpaceX has proposed that LEO satellite opera-

---

[68] See Article 21 of the Radio Regulations.

[69] WRC-15, Report of the Director on the Activities of the Radiocommunication Sector, Part 2, 2015, para.3.2.2.4.3.

[70] WRC-15, Report of the Director on the Activities of the Radiocommunication Sector, Part 2, 2015, para.3.2.2.4.3.

[71] D. St. John, The Trouble with Westphalia in Space: The State-Centric Liability Regime, Denver Journal of International Law and Policy, Vol. 40, 2012, p. 688.

[72] J. Zoller, Improving the international satellite regulatory framework, 2011.

tors share data to indicate the steering angle of each beam within a satellite's footprint to decrease the occurrence of inline events with geostationary satellites.[73] OneWeb committed to coordinate in good faith with the Radio Astronomy Service in case of interference.[74] OneWeb and Boeing even went as far as concluding an orbiting settlement to regulate the distance at which their respective constellations should orbit (both supposed to orbit at around 1200 km in the LEO).[75]

Because operators expressed their clear intention to cooperate to achieve coordination, it appears reasonable to wait to see whether cooperation and coordination will be used by operators to solve their interference problem, hence suppressing the need for any additional regulation. Ultimately, the solution to interfering, conflicting, or colliding satellite fleets might well be technological and not regulatory.

4.5.4   The Answer Might Be Technological and Not Regulatory

Indeed, technological developments might suppress future interferences between constellations and thereby the need for regulation of those LEO fleets. OneWeb has a patent pending on progressive pitch technology which would allegedly provide "a technique to avoid interference between their low-earth orbit constellation and geostationary satellites" so that "their satellites will automatically change orientation and power level as they pass over the equator to avoid interference with geostationary satellites orbiting above them."[76] This technique reminds of the non-operating zone around the geostationary arc system used by first-generation LEO constellation SkyBridge to maintain the interference level with geostationary satellites at a nonvisible level: to avoid creating interference with geostationary systems, those satellites were programed with a non-operating zone covering 10° on both sides of the GEO arc as seen from the ground, calculated to take into account the antenna directivity and the different power levels.[77]

Further, exciting technological progresses might be made in the field of optical communication satellites and ground systems: by relying on lasers, rather than on

---

[73] Consolidated Reply of Space Exploration Holdings, LLC before the FCC, July 14, 2017, p. 4.

[74] Opposition and Response of Worlvu Satellites Limited before the FCC, August 25, 2016, para. E, p. 11.

[75] *See* P. de Selding, OneWeb, Boeing settle constellation orbit issue; SpaceX questions OneWeb ownership, Space Intel Report blog entry, Apr 25, 2017, available at: https://www.spaceintelreport. com/oneweb-boeing-settle-constellation-orbit-issue-spacex-questions-oneweb-ownership/ (link accessed on September 30, 2018).

[76] Opposition and Response of Worlvu Satellites Limited before the FCC, August 25, 2016, Annex A-1; *See* also L. Press, Inevitability of global standards for non-terrestrial spectrum sharing, Blog entry, 2017, available at: http://www.circleid.com/posts/20171inevitability_of_global_standards_ for_non_terrestrial_spectrum_sharing/ (lastly checked September 30, 2018).

[77] P. Sourisse, D. Rouffet and H. Sorre, SkyBridge: a broadband access system using a constellation of LEO satellites, ITU Press Archiv WRC97, available at: https://www.itu.int/newsarchive/press/ WRC97/SkyBridge.html (accessed on September 30, 2018).

radio waves, to send and receive data, those systems would be able to avoid the potential interference and latency inherent to crowded radio-frequency spectrums.[78]

Given the fact that all past projects of first-generation constellations failed, it might be wise to wait to see how the deployment of the new massive satellite fleets go before developing specific rules to regulate them and to remember the wise word of Judge Lachs[79]:

> [Lawmaking] is a long and painstaking process: it is a continuous process in which the lawmakers must remain watchful; facing the existing and changing requirements of life. We have to welcome what has been achieved and strive for further agreements. The law of the outer space is in its formative stage only. We must proceed with prudence and care – take the full benefit of agreements reached, work on them, extend them, make them a living reality and continue with our efforts for further agreements.

Alice Rivière LL.M. is admitted as an attorney-at-law in the State of New York and as *avocat à la cour* in Paris. She holds an LL.M. in International Law from the University of Miami (2015), a Master's degree in International Economic Law from Université Paris I Panthéon-Sorbonne/ Columbia Law School/Sciences Po Paris (2012), a Maîtrise en droit from Université Paris I Panthéon-Sorbonne, and an LL.M./Magister legum from Universität zu Köln. After practicing law in the field of international dispute settlement in Switzerland for a couple of years, she recently joined the space industry and started as Anticorruption Compliance Manager with Airbus Defence and Space in Munich, Germany.

---

[78] C. Anderson and K. Taffelson, SpaceX, STARLINK and the battle for satellite broadband development, Space Angels blog post, 2017, available at: https://www.spaceangels.com/post/spacex-starlink-and-the-battle-for-satellite-broadband-deployment (lastly check on September 30, 2018).

[79] Judge Manfred Lachs, Additional Report of the Committee on the Peaceful Uses of Outer Space, A/5549/Add:1, Verbatim record of the Twenty-fourth Meeting, Annex, p. 4.

Chapter 5
The Future Impact of the ITU Regulatory Framework on Large Constellations of Satellites

Claudiu Mihai Tăiatu

Abstract This research will address the future impact of the ITU Radio Regulations and ITU-R Recommendation on large constellations of satellites. In particular, this research will analyze the latest findings related to the need for an improved ITU regulatory framework that will specifically address large constellations of satellites.

This research will bring clarification to the problem of bringing into use of frequency assignments to non-GSO systems by analyzing the work of the ITU-R about the implementation of a milestone-based approach system for the future deployment of large constellations of satellites in orbit. In this regard, this research will analyze the necessary legal framework proposed by the WP4A to be discussed at the WRC-19 to prevent the risk of speculative fillings and spectrum warehousing and avoid administrations to block spectrum for others. This research will also address the need for a new ITU-R Recommendation addressing the risk of space debris posed by large constellations of satellites in LEO. Reference will be made to mitigation and remediation of space debris from the ITU's perspective. The role of the ITU will be analyzed also by referring to the *Guidelines for the long-term sustainability of outer space activities*.

5.1 Introduction

The International Telecommunication Union (ITU) is the specialized agency of the United Nations responsible for the allocation of global radio spectrum and satellite orbits in accordance with principles of equitable access, rational use, and protection

C. M. Tăiatu (✉)
International Institute of Air and Space Law (IIASL), Leiden University,
Leiden, The Netherlands

© Springer Nature Switzerland AG 2019
A. Froehlich (ed.), *Legal Aspects Around Satellite Constellations*, Studies in
Space Policy 19, https://doi.org/10.1007/978-3-030-06028-2_5

from technical harmful interference.[1] The radio frequency spectrum is a limited natural resource that must be used rationally, efficiently, and economically.[2]

For the ITU, the deployment of large constellations of satellite poses two main concerns.[3] The frequency assignments and orbital slots are assigned to the requesting administrations by a complex system of notification, coordination, and bringing into use (BIU), but current provisions of the ITU Radio Regulations (RR) do not specifically address BIU of frequency assignments to non-GSO satellite systems and may not be sufficient to prevent the risk of spectrum warehousing and paper satellites. In addition, large constellations of satellites could potentially affect the environmental sustainability of outer space, meaning that mitigation and remediation of space debris is important for the scope of the ITU in relation to the advent of commercial activities in LEO.[4]

The imminent deployment of large constellations of satellites is challenging the ITU legal framework. The research question is related to the need for an improved ITU regulatory framework specifically addressing large constellations of satellites. In particular, the research questions are referring to the following: (1) what is the perspective for BIU frequency assignments to non-GSO satellite networks, and (2) how should the ITU become active on mitigation and remediation of space debris specifically to LEO?

5.2 For the Purpose of Securing Rights and Protection from Technical Harmful Interference of Subsequent Networks, the BIU of Frequency Assignments Is Mandatory

The implementation of new space radiocommunication systems involves national administrations to notify satellite frequency assignments and associated orbital positions for recording in the Master International Frequency Register (MIFR), declaration of Bringing into Use (BIU) of frequency assignments to a satellite network, and protection of these assignments from harmful interference.[5] The RR

---

[1] According to Article 2 (a) of the ITU Constitution, the ITU shall: *effect allocation of bands of the radio-frequency spectrum, the allotment of radio frequencies and the registration of radio-frequency assignments and, for space services, of any associated orbital position in the geostationary-satellite orbit or of any associated characteristics of satellites in other orbits, in order to avoid harmful interference between radio stations of different countries.*

[2] ITU Constitution, Chapter VII, Article 44 *Use of the Radio-Frequency Spectrum and of the Geostationary-Satellite and Other Satellite Orbits.*

[3] Lyall F, Larsen P.B., *Space Law, A treatise,* (Chapter 9-*Unusual Problems,* p. 240, ISBN: 978-1-4724-4782-1, 2nd Edition, Routledge, 2018).

[4] Ibid. *supra* note 3, p. 244.

[5] Matas Attila, "The ITU Space Regulation – A Key Element to Access Space", [2018] IAC-18. E7.7-B3.8.6.x42748.

contains the relevant regulatory principles and coordination procedures for the use of frequencies and for avoiding harmful interference with regard to other existing and planned terrestrial or space service users. Three basic steps involve the following[6]:

- advance publication of information (API)—Section I, Article 9 RR;
- coordination (CR/R)—Section II, Article 9 RR;
- notification and recording of frequency assignments in the MIFR—Article 11 RR.

The BIU of frequency assignments to a satellite network allows satellite services to be operated in specific frequency bands and will lead to a permanent registration of the frequency assignments in the MIFR.[7] The ITU Radiocommunication Bureau (BR) manages the MIFR for information purposes.[8] The BIU of frequency assignments is addressed in Article 11 RR, Section II. According to the provisions of No.11.44 RR, the time limit for BIU frequency assignments in nonplanned bands is 7 years following the date of receipt of the relevant complete information received according to Article 9 RR. During the 7-year regulatory period, administrations have to complete the procedure for API[9] or CR/R[10] and provide the date of bringing into use (DBIU).

The provisions of Article 11.44B RR complement the requirements for BIU by providing the conditions under which a frequency assignment is considered BIU.[11] The first condition is that *at least one space station with the confirmed capability of transmitting or receiving that frequency assignment* has to be deployed in orbit. This means that at least one satellite has been deployed in orbit and can provide service. The second condition imposes a continuous period of 90 days in orbit:

---

[6] Ibid. *supra* note 5.

[7] Matas Attila, Henri Yvon and Loo Chuen Chen, "The ITU Radio Regulations Related to Small Satellites" pp. 237–264, in Marboe Irmgard (ed), *Small satellites: Regulatory challenges and chances* (Lam edition, Brill Nijhoff, 2016).

[8] ITU Addendum 1 to Document CMR15/ 4-E, *Report of the Director on the Activities of the Radiocommunication Sector*, 2 July 2015, pp. 6/75 in Chapter 2. Application of the Radio regulations for Space Services, 2.1. Introduction. <https://webcache.googleusercontent.com/search? q=cache:wJ8OZvvEM-kJ:https://www.itu.int/md/dologin_md.asp%3Flang%3Den% 26id%3DR15-WRC15-C-0004%2521A1%2521MSW-E+&cd=1&hl=ro&ct=clnk&gl=ro> accessed: 03.06.2018.

[9] Article 9, No. 9.1 or 9.2, Chapter III, RR.

[10] Article 9, No. 9.1A, Chapter III, RR.

[11] No.11.44B RR "*A frequency assignment to a space station in the geostationary-satellite orbit shall be considered as having been brought into use when a space station in the geostationary-satellite orbit with the capability of transmitting or receiving that frequency assignment has been deployed and maintained at the notified orbital position for a continuous period of 90 days. The notifying administration shall so inform the Bureau within 30 days from the end of the 90-day period. On receipt of the information sent under this provision, the Bureau shall make that information available on the ITU website as soon as possible and shall publish it in the BR IFIC. Resolution 40 (WRC-15) shall apply.*"

deployed and maintained at the notified orbital position for a continuous period of
90 days.

5.3   The Provisions of the ITU Radio Regulations (RR) Do Not Specifically Address BIU of Frequency Assignments to Space Stations in Non-GSO Satellite Systems

The practice of the BR for non-GSO is reflected in the Rules of Procedure for RR No. 11.44 stating that a frequency assignment to space stations in any non-GSO system has been brought into use when one satellite from a planned system is deployed in space and capable of transmitting and/or receiving that assignment, irrespective of the number of satellites.[12]

Basically, the current issue is that with only one satellite recorded in orbit, the administrations could be treated as if the satellite operators would have deployed all satellites of the non-GSO network, meaning that with only one satellite, the administrations will receive international recognition and protection for the recorded frequency assignments irrespective of the number of satellites the constellation is composed of.

The lack of specific provisions in the RR is discussed by the BR ITU-R Study Group 4A (WP4A), which supports studies related to the "Efficient orbit/spectrum utilization for FSS and BSS"[13] under WRC-19 Agenda Item 7.[14] The WP4A is responsible for WRC-19 Agenda Item 7 Issue A referring to "Factors related to the bringing into use of a frequency assignments of non-GSO systems subject to coordination."[15] In this role, the WP4A analyzes a future mechanism to be proposed at the WRC-19 for BIU frequency assignments to non-GSO constellations, for purposes of the application of Article 11RR.[16] The results of these studies are reflected in the Draft CPM Report for WRC-19 Agenda Item 7—Issue A, which will be analyzed below.

---

[12] Director, ITU Radiocommunication Bureau, *Report of the CPM on operational and regulatory/procedural matters. To the World Radiocommunication Conference 2015*, (Chapter 5, Issue G, page 48, 2nd Session of the Conference Preparatory Meeting for WRC-15).

[13] Document CPM19-1/1-E, "Director Radiocommunication Bureau – Structure of ITU-R Study Groups", 24 November 2015.

[14] WRC-19 Agenda Item 7: "*to consider possible changes, and other options, in response to **Resolution 86** (Rev. Marrakesh, 2002) of the Plenipotentiary Conference, an advance publication, coordination, notification and recording procedures for frequency assignments pertaining to satellite networks, in accordance with **Resolution 86** (Rev.WRC-07), in order to facilitate rational, efficient and economical use of radio frequencies and any associated orbits, including the geostationary satellite orbit*".

[15] Annex 22 to Document 4A/63-E.

[16] Art. 11 of the Radio Regulations ITU edition 2016, regarding the Notification and recording of frequency assignments, WRC 2015.

5.4   The *Draft* Conference Preparatory Meeting (CPM) for WRC-19: Report 2018

The Draft CPM Report on technical, operational and regulatory/procedural matters to be considered by the 2019 World Radiocommunication Conference was issued on September 17, 2018, by the ITU Director of the Radiocommunication Bureau.[17] This is the latest document containing the work of the responsible ITU-R groups involved in the preparation for the WRC-19, including the work of the WP4A on the agenda Item 7 Issue A presented in the Draft CPM Report under the title "Bringing into use of frequency assignments to all non-GSO systems, and consideration of a milestone-based approach for the deployment of non-GSO systems in specific frequency bands and services."[18]

(a)   The first highlight of the Report is the guiding principles that were developed to assist in resolving Issue A under WRC-19 Agenda Item 7:

> 1) The BIU process should be separate from any follow-up actions required to maintain the rights and protections for the recorded frequency assignments to non-GSO systems.
> 2) The successful completion of the BIU process for non-GSO systems does not require the deployment of all satellites in the system by the end of the seven-year regulatory period.
> 3) Appropriate time should be given to allow the completion of the deployment of non-GSO systems.
> 4) Appropriate transitional measures should be considered to address the implications of any new milestones adopted by WRC-19.
> 5) The milestone-based approach should be applied to all non-GSO systems in specific space services in specific frequency bands.
> 6) Concurrently with the development of a milestone-based approach, methodologies should be developed for the implementation of RR Nos. 9.58, 11.43A, and 11.43B relating to the regulatory treatment of the adjustments to the characteristics of frequency assignments to non-GSO systems.
> 7) The milestone-based approach should provide incentives to notifying administrations to deploy satellites in a timely manner, as a failure to meet a given milestone for a non-GSO system will result in consequences.[19]

These principles reflect the work of the WP4A and represent the basis for the regulatory options proposed for BIU frequency assignments to non-GSO satellite networks. Among the keywords reflected by these principles is the need to understand the deployment of satellites for BIU purposes as a prerequisite for maintaining the rights and protections of the recorded frequency assignments that will ultimately be reflected in a milestone-based approach; to develop appropriate transitional measures to address the implications, including legal, for any new milestones; and also to address the possibility of adjustments in the MIFR in case of failing to meet a

---

[17] Director, ITU Radiocommunication Bureau—Document CPM19-2/1-E, "Draft CPM Report, Conference Preparatory Meeting for WRC-19, Geneva, 18-28 February 2019" (Plenary Meeting, 17 September 2018).

[18] Ibid. *supra* note 17, p. 28.

[19] Ibid. *supra* note 17, p. 30.

given milestone. Based on the guiding principles mentioned above, two general conclusions could be identified in the draft CPM text in the executive summary.

The first general conclusion refers to the fact that the 7-year regulatory period should continue to be applied to BIU frequency assignments to non-GSO, and three options were proposed in regard to the minimum continuous period for which a satellite should be maintained in a notified orbital plane: (1) 90 days, (2) less than 90 days, (3) no fixed period.[20] The second general conclusion is that a new WRC Resolution should be adopted to implement a milestone-based approach for the deployment of non-GSO systems, and examples of regulatory implementation were proposed in the draft CPM Report. Delegations at the ITU proposed several options regarding the number of milestones, periods, percentage for each milestone, and consequences of failing to meet a milestone.

(b) The second highlight of the Report is the regulatory solutions proposed to be adopted by the WRC-19 for Agenda Item 7 Issue A. The draft new Resolution [A7 (A)-NGSO-milestones] (WRC-19) titled "A milestone-based approach for the implementation of frequency assignments to space stations in a non-geostationary-orbit satellite system in certain frequency bands and services" implies modification of the ITU Radio Regulations to reference draft new Resolution into Article 11. Also, modification to Article 13 RR was mentioned. In addition, the draft new Resolution contains in Annex 1 the proposed information to be submitted in the deployment plan while in Annex 2 the method of calculation of the milestone periods.

Analyzing the content of the draft new Resolution, it could be noticed several frequency bands and services, transition alternatives, modification information, including consequences of nonsubmission and suspension. The frequency bands and services considered for the application of the milestone-based approach are divided into bands generally and not generally agreed for inclusion. The instructions to the Radiocommunication Bureau (BR) are not yet included but only mentioned.

5.5   The BIU of Frequency Assignments to Non-GSO Satellite Constellations Is a Prerequisite to Any Milestone-Based Approach System

In the perspective of the WRC-19, the WP4A proposed that the BIU of frequency assignments to non-GSO systems should continue to be achieved for all frequency assignments to non-GSO systems, with the deployment of a single satellite into a

---

[20] Ibid. *supra* note 17, p. 28.

notified orbital plane, within 7 years of the date of receipt of the API or request for coordination CR/C, as applicable.[21]

This means that regardless of the number forming the future LEO satellite constellation, the 7-year regulatory period for BIU of frequency assignments to GSO satellites should be maintained for non-GSO satellite systems. Additionally, the rule that at least one satellite capable of transmitting or receiving the frequency assignments has been deployed in a notified orbital plane should apply equally to all non-GSO systems—whether to a single-satellite non-GSO system/network or to a multiplane, multisatellite constellation of non-GSO satellites.[22]

The procedure for BIU of frequency assignments to a non-GSO satellite system is presented as a prerequisite for securing rights and protections for the frequency assignments for the entire non-GSO satellite system. The accomplishment of BIU frequency assignments to non-GSO systems will provide the administrations the rights and protections for the recorded frequency assignments as if the notified constellation will be on orbit.

This approach is considering BIU as a preliminary step to the milestone-based approach system and a prerequisite for securing rights and protections for the frequency assignments previously recorded in the MIFR. Thus, the administrations will not be conditioned by the capacity limitations for the launch of the satellites provided by satellite operators when securing the rights and protections for frequency assignments. In addition, frequency assignments would be secured and protected on the basis of priority. Timely registrations at the ITU would reward administrations that posed the effort to register their project constellations at the ITU.

5.6 The Milestone-Based Deployment Approach Is Necessary for Preserving in the MIFR the Recording of the Frequency Assignments to Non-GSO Systems in Specific Bands and Services

The milestone-based approach is a process used to assess that the use of frequency assignments to non-geostationary systems is compliant with their associated notified characteristics. The number of satellites deployed into one or more notified orbital planes, with the confirmed capability of transmitting or receiving the frequency assignments, will be compared with the minimum number of satellites

---

[21] Document ITU 4A/TEMP/188, "Compilation Preliminary Draft CPM Text for WRC-19 Agenda Item 7 – Issue A", [2018], Radiocommunication Study Groups.

See also the latest document: *supra* note 17—Document CPM19-2/1-E, "Draft CPM Report, Conference Preparatory Meeting for WRC-19, Geneva, 18-28 February 2019" (Plenary Meeting, 17 September 2018).

[22] Ibid. *supra* note 21.

required as per the milestone.[23] The milestone-based approach is the solution identified by WP4A that allows establishing a balance between the need to prevent spectrum warehousing and the need to recognize the technical and operational challenges associated with this type of non-GSO system, including the proper functioning of coordination mechanisms.

The successful completion of the BIU of non-GSO satellite systems should not require the deployment of all satellites in the recorded system by the end of the 7-year regulatory period. The controlled deployment of the non-GSO systems will happen only beyond the 7-year regulatory period for BIU according to milestones.

The retention of the rights and protections for the recorded frequency assignments is not assured unless the recorded frequency assignments of the large non-GSO satellite system reflect its deployment after the end of the 7-year regulatory period.

The advantage is to secure and protect the frequencies even if the satellites are not yet in orbit, but the retention of rights and protections for the recorded frequency assignments would be assessed in accordance with the conditions (duration and percentage) established by a milestone-based approach.

The administrations will have to notify the ITU for the BIU of frequency assignments for non-GSO satellite constellations. After completing this step, satellite operators would continue the deployment of the constellation according to the rules to be developed for the milestone-based approach system.[24]

If provisions will be included into a new WRC Resolution, it would allow developing a solution for the non-GSO system implementation issue without having to change a large number of provisions in the RR at the WRC-19, and if the approach is successful, it can eventually be built into Article 9 RR and/or Article 11 RR. The earliest reference date for the application of a milestone-based approach could be the date of entry into force of the Final Acts of the WRC-19.

It is worth mentioning that none of the non-GSO satellite networks have been BIU. Microsat-2a and Microsat-2b developed by SpaceX as part of its Starlink constellation were put into orbit in 2018 for the purpose of testing spacecraft and antenna technology. OneWeb has instead delayed several times the deployment of its satellites, last time for February 2019. For administrations that filled in 2014, the deadline for BIU is 2021 according to the 7-year regulatory framework.

---

[23] Ibid. *supra* note 21.

[24] Ibid. *supra* note 21.

5.7   Three Milestones with a Total Timing of Twelve Years and a Final Percentage of Seventy-Five of the Number of Recorded of Assignments to Non-GSO Systems in Specific Bands and Services

A three milestone-based approach with a total of 12 years (7 years regulatory + 7 years deployment milestones) was proposed to date.[25] The three milestones will be applied from the end of the 7-year regulatory period, which is after (1) the date of the receipt of API or (2) the coordination request information, as applicable to non-GSO systems filed prior to or after July 2016. More options are available for the WRC-19 consideration in regard to a milestone approach (see Table 5.1), but the overall deployment under three milestones and a time frame of 12 years should encourage quick deployments of non-GSO systems and avoid the operational constraints of protecting an entire filing, which could consist of hundreds or even thousands of "paper" satellites, but where, in reality, a single non-GSO spacecraft is actually in service.

It could be reasonably affirmed that the intent of the administrations and the satellite operators is to deploy all satellites. The reality is that in the process of deployment, operators may lose some satellites. This is the reason why the last milestone cannot be established at 100% without the risk of being challenged by the BR. If considering 100%, in case of losing a satellite or number of satellites, the administrations may not be subject to compliance with the filling and replacement may not happen overnight. A 75% for the last milestone will allow replenishment or other technical problems to be solved and would eliminate the risk to make modifications in the MIFR at the last milestone.

One of the most debated aspects is the percentage to be decided for the last milestone. If the last milestone will be 75%, the percentage will benefit the satellite operators and will also be sufficient to assure real operations and avoid blocking spectrum. If the last milestone would be established at 100% and the operator would lose a satellite, the administration could be challenged by the BR for not complying with the filling in the MIFR. Instead, the 75% will not penalize the operators and will give them flexibility to make the deployment. But one of the challenges for the final milestone with 75% is that it will leave the door open for the deployment of the notified constellation, and it will become a burden in case of coordination to avoid harmful interference. If the final configuration will be decided to be 100%, operators must be very careful about having the satellites available for operations. The final milestone is disputable, and only the WRC-19 could clarify the solution.

---

[25] Ibid. *supra* note 21.

See also: Annex 6 to Document 4A/364-E, *"Working Document Towards a Preliminary Draft New Report ITU-R S – NGSO_FSS_BIU"*, 18 May 2017 and temporary Report 4A/TEMP/123-E issued 11 May 2017. Document 4A/364-E, 9 June 2017, Chairman, Working Party 4A, Report on the Meeting of WP4A 3-12 May; revised February 2018.

Table 5.1 Options for the milestone-based approach[a]

Milestone timing[b] Number of years after the end of the 7-year regulatory period			Minimum required percentage of satellites to be deployed to meet the milestone[c]	Deployment Factor	
M1		1		A1 & F1: 10%		10
		2		B1: 8.33%		12
			P1	C1 & D1: 10%	DF1	10
				G1: 30%		3.33
		4		E1: 10%		10
M2		3		A2 & F2: 33%		3.03
		4		C2: 30%		3.33
			P2	B2: 25%	DF2	4
		5		D2: 50%		2
		7		E2: 75%		1.34
		2+A[d, e]		G2: 60%		1.66
M3		5		A3: 75%		1.34
		6		B3: 75%		1.34
			P3	F3: 100%	DF3	1
		7		C3: 90%		1.11
				D3: 100%		1
		8		E3: 100%		1
		2+A+B[f]		G3: 100%		1

[a]Options for the milestone-based approach. Table 3/7/1.3.2-1 in Document CPM19-2/1-E, "Draft CPM Report, Conference Preparatory Meeting for WRC-19, Geneva, 18-28 February 2019" (Plenary Meeting, 2018)

[b]Initial timing is the date of receipt by the Bureau of the relevant complete information under RR No. **9.1** or No. **9.1A** of Article **9**, as appropriate

[c]In this column, (A1, A2, A3) (B1, B2, B3), (C1, C2, C3), (D1, D2, D3), (E1, E2, E3), (F1, F2, F3) and (G1, G2, G3) represents all the combinations of three milestones identified in the studies for the implementation of the milestone-based approach. For all options, except option (F1, F2, F3), the date for commencement of the milestone process based on the end of the 7-year regulatory period is 1 January 2021. For Option (D1, D2, D3 & F1, F2, F3) an alternative date of commencement is 23 November 2019. For option (B1, B2, B3), it could be a date between 1 January 2021 to 1 January 2024, and may be considered for other options, as the case may be

[d]A & B are variables: 12 months $\leq$ A, B $\leq$ 30 months based on the conditions met

[e]A = (number of satellites launched/30% of the total number of satellites in the MIFR) *30. Where the resulting number calculated should be rounded up to the greatest whole number

[f]B = (number of satellites launched/60% of the total number of satellites in the MIFR) *30. Where the resulting number calculated should be rounded up to the greatest whole number

Another aspect to be discussed is the period of the first milestone. This is the most important milestone because operators do not have experience with large constellations of satellites and also have to assure coordination. It could be reasonably mentioned that 4 years after the regulatory period, meaning 7 + 4 years (11 years), would not be acceptable to many administrations. If such solution would be decided, administrations could block spectrum for 11 years with only one satellite deployed, without the satellites deployed to be in operation, and this is not the scope of the milestones. This is the reason why this option seems to be the most unlikely.

5.8   The Deployment Factor Should Ensure That the Total Number of Satellites That Can Be Deployed Is Adjusted According to the Progress Made by Each Milestone

The deployment factor is part of the WP4A efforts to find a solution that would prevent administrations to lose their rights and protections in case of failing to meet the percentage of deployment imposed by the milestones.

The deployment factor is intended to determine the maximum number to be referred to in the MIFR in case of noncompliance with a milestone.

This approach in relation with noncompliance with the milestones would assure the adjustment in the MIFR. Deployment factor is provided in order to scale up the constellation deployment. In this case, the maximum number of satellites is obtained by multiplying the number of satellites actually deployed and declared by the notifying administration by the deployment factor.[26]

Any system that meets its milestone implementation will be entitled to the recognition of its notified frequency assignments and continue to have those assignments recorded without change in the MIFR. The progress of each constellation has to be reflected in the MIFR according to the real deployment of satellites inside the milestone-based approach.

If a milestone is not met, the result would be an adjustment of the MIFR that would align the MIFR for the subject frequency assignments with the numbers of planes and satellites per plane deployed into the system at the expiry of the deadline for the milestone in question.[27]

It could be reasonably stated that without the deployment factor approach, the provisions of RR No. 13.6 will have consequences toward the modification of frequency assignments and satellites registered during the coordination process.

The maximum number of satellites you can deploy is registered in the MIFR; deployment factor depends on this number. The satellites deployed in orbit must be reflected in the MIFR; a close alignment between the MIRF and real satellite deployment in orbit is needed.

The deployment factor only helps when the deployment is close to the milestone requirements; otherwise, the satellites will fly with no purpose.

---

[26] Ibid. *supra* note 21.

[27] Ibid. *supra* note 21.

5.9 Transitional Measures Should Be Considered for the Non-GSO Satellite Systems Brought Into Use Prior to the WRC-19

Another point to be considered by the decision makers at the WRC-19 is the predictability of the law and the nonretroactivity principle. According to the WP4A findings, transitional measures represent the legal basis for imposing a milestone-based approach system as a condition for the retention of rights and protections gained through the BIU notification, considering that a milestone-based approach system was not in place at the moment of BIU prior to the WRC-19.[28]

The implementation of transitional measures will assure equitable bringing into use for all the non-GSO satellite systems. Transitional measures are needed to address the deployment of non-GSO systems for which frequency assignments were brought into use prior to the WRC-19 but which have fewer satellites in orbit than the notified number of satellites per orbital plane deployed before the end of the WRC-19. It could be reasonably mentioned that in case some operators will BIU the frequency assignments before WRC19, not all the notified satellites will, however, be deployed. Only this situation triggers the need for transitional measures.

It could be stated that the fundamentals of the transitional measures is to assure that (1) administrations should not undergo a bringing into use process a second time and (2) a milestone-based approach system would be applicable to such systems in the context that at the moment of BIU, the only condition that had to be respected was fulfilled by the administration.

Another challenge that needs to be addressed through transitional measures is to provide that under a milestone-based approach system, the noncompliance could lead to adjustments to the characteristics of the recorded assignments in the MIFR according to the real deployment of satellites. Without this clarification, under the provisions of RR No.13.6, any interested administration could challenge the rights and protections acquired under the current BIU procedure.

Transitional arrangements could depend on the characteristics of the milestone-based approach adopted by the WRC-19, a reason why they would be necessary only after such a decision will be made. It was proposed by the WP4A that options for transitional measures should start at the end of the WRC-19 and some variables should be considered for different situations. Transitional measures are more than one for different case scenarios.[29]

Two options have been identified in the Draft CPM Report 2018 for transitional measures: one applying identical milestones, associated timelines, and required level of deployment with a reference point for the commencement of the milestones period,[30] while the second option would consist of different sets of milestones

---

[28] Ibid. *supra* note 21.

[29] Ibid. *supra* note 21.

[30] 3/7/1.3.2.2.1 – *Option 1, (…) For non-GSO systems with frequency assignments reaching the end of their seven-year regulatory period after a date to be set by the Conference, the commencement*

depending on whether or not the 7-year regulatory period was reached prior to the entry into force of the future WRC-19 decision on Agenda Item 7 Issue A.[31] The decision to adopt one of the two options is a matter of policy and would also depend on the characteristics of the milestone-based approach to be adopted by the WRC-19.

5.10   The Condition of Continuous Operation in a Notified Orbital Plane with the Capability of Transmitting or Receiving the Frequency Assignments

The WRC-19 should clarify whether the 90 days of continuous operation applies or not for BIU frequency assignments to non-GSO.

In principle, the 90-day continuous operation was necessary to prevent satellite hopping, which would not be a problem for non-GSO as it is for GSO.

On the other hand, the lack of a continuous period for non-GSO could advantage orbital planes for 1 day or a similar short period, for instance the use of a single non-GSO satellite to BIU of frequency assignments with various filings at different orbital altitudes, after launch in a lower orbit and before the final intended orbit is reached, even if administrations have no intention to operate in such intermediate orbits.

Reasonably, if the continuous period will be decided as relevant, then different approaches could be implemented depending on the type of non-GSO systems. Current discussions at WP4A have resulted in three options, which are contained in the Draft CPM Report 2018 and prepared to be presented to the WRC-19 for decision (see Table 5.2).

---

of the milestone period will be the actual date of the end of the seven-year regulatory period. For the non-GSO systems with a regulatory period that ends before the date to be set by the Conference, the commencement of the milestone process is based on that date. Options studied for the date to be set include 23 November 2019 (the first day after the end of the conference), 1 January 2021 and 1 January 2024.

[31] *3/7/1.3.2.2.2 – Option 2, (…) not only would there be a different reference point for the commencement of the milestone-based approach but the actual approach (i.e. associated timelines) would be different and would depend upon the date of the end of their seven-year regulatory period. Extra time would be granted to non-GSO systems for which the end of the seven-year regulatory period comes before the date of the commencement of the regular milestone-based approach (…).*

Table 5.2 Options relating to the continuous period for confirming BIU[a]

Options	Descriptions
A	A continuous period of at least 90 days in a notified orbital plane of a satellite with the capability of transmitting or receiving the frequency assignments. *Applicable to some non-GSO systems based on RoP on RR No. 11.44 (Ed. of 2017)*
B	A continuous period of X (1–90 days) of deployment in a notified orbital plane of a satellite with the capability of transmitting or receiving the frequency assignments may be sufficient. *The 90-day duration may not be required for the non-GSO administration/operator to determine that a space station with the capability has been deployed in a notified orbital plane*
C	No fixed period. *Administration informs the Bureau of BIU once it confirms deployment of a space station with the capability of transmitting/receiving the frequency assignments into one of the notified orbital planes*[b]

[a]Options relating to the continuous period for confirming BIU. Table 3/7/1.3.1-1 in *supra* note 18
[b]The studies have shown that for some services, e.g., the radio navigation-satellite service, no fixed period is required. Instead, the administration/operator requires only as long as it takes to confirm the deployment into a notified orbital plane of a satellite with the capability of transmitting or receiving the frequency assignments. This can vary from system to system, but will not require 90 or more continuous days of deployment. For this reason, no fixed continuous period is required for these particular systems

5.11 Maintaining the Constellation After BIU

The satellites in LEO have only few years of lifetime, which could create certain difficulty for the operators to manage a constant number of satellites in the constellation and replace them with regularity. In this regard, discussions about the milestones and deployment of satellites could be extended post BIU. This includes understanding of the procedure for suspension of the satellite systems. It is not yet clear whether the mechanism of milestones will be used in case of bringing back into use (BBIU) and how the procedure will respect the MIFR notification.

It could be reasonably mentioned that once a milestone-based approach system is imposed for the retention of the rights and not for BIU itself, the current procedure for the suspension and bringing back into use frequency assignments needs to be modified. The provisions of Article 11, No. 11.49, and particularly 11.49.1 of the RR are not suitable for a milestone-based system. To bring back into use, the same percentage that was used to retain the rights and protections would have to be respected; otherwise, the frequencies will be suppressed. In this case, the suspension of constellation could intervene only after complying with all the milestones and the real satellites in orbit will correspond with the MIFR registration.

Market access and costs of service represent a secondary but important step after the BIU frequency assignments to non-GSO satellite systems and will be further explained. Also, if the large constellations of satellites will become a success, it is important to discuss about equitable access, in particular sharing of spectrum and coordination for protection from harmful interference. The function of splitting spectrum to various users will have to be made through frequency management at national level. The priority is also avoiding in orbit collisions between the satellite systems of the large constellations of satellites, and this could represent another form of coordination at national level.

5.12 National Licensing—The Case of U.S.: SpaceX and OneWeb

The regulatory responsibility for national radio spectrum is assigned in the U.S. between the Federal Communications Commission (FCC) and the National Telecommunications and Information Administration (NTIA). For non-Federal use, meaning commercial and private business, the FCC can provide a license and allocate the spectrum to be used nationally.[32]

Back in June 2017, the FCC has granted OneWeb access to the U.S. market for its broadband satellite constellation. This was the first license of its kind globally. OneWeb has also received an extension for more satellites. The FCC granted approval to OneWeb to provide services in the U.S. market, but even if OneWeb was firstly mentioned, it did not acquire priority.[33]

The FCC has authorized in March 2018 the Space Exploration Holdings, LLC (SpaceX), application from November 2016 to construct, deploy, and operate a proposed nongeostationary orbit (NGSO) satellite system comprising of 4425 satellites for fixed-satellite service (FSS) around the world.[34] The authorization mentions that this approval serves the public interest and will allow SpaceX to bring high-speed, reliable, and affordable broadband service to consumers in the U.S. and around the world.[35]

OneWeb has requested SpaceX application to be conditioned by a safety buffer zone that would prevent collisions between the satellites of the two operators. The FCC advised the operators to reach an interoperator coordination agreement and coordinate physical operations at similar orbital altitudes. Only if such agreement will not be reached that the Commission will intervene and may impose requirements that would involve a wide range of changes in system design and operations.[36] This results in an important national policy that directs efforts to interoperator coordination agreements. These agreements are complementary to the ITU coordination, which is an international obligation. The difference is that the ITU coordina-

---

[32] See https://www.fcc.gov/engineering-technology/policy-and-rules-division/general/radio-spectrum-allocation.

[33] FCC 17-77, 2017, Order and Declaratory Ruling, OneWeb NGSO FSS System.

[34] FCC 18-38, 2018, IBFS File No. SAT-LOA-20161115-00118—Application for Approval for Orbital Deployment and Operating Authority for the SpaceX NGSO Satellite System. <https://www.fcc.gov/document/fcc-authorizes-spacex-provide-broadband-satellite-services> accessed 21.09.2018.

[35] Part III, paragraph 7.

[36] Part III, chapter 11.

tion refers to avoidance of harmful interference, while the coordination imposed by the FCC refers to space debris.

In addition, SpaceX answered the inquiry of the FCC regarding mitigation of space debris and stated that it will aim for a 1-year post-mission disposal with perigee of 300 km to avoid disruption of ISS activities if still in orbit. The first major post-mission disposal (PMD) will be around 2025.[37]

The FCC issued the Report and Order and Further Notice of Proposed Rulemaking in September 2017, by which it adopted regulatory modifications to update Parts 2 and 25 concerning Non-Geostationary, Fixed-Satellite Service Systems and Related Matters, initiated in 2016 with the Notice of Proposed Rulemaking.[38] The modifications were well received, and the majority came into effect from January 17, 2018. One of the modifications refers to NGSO milestones, imposing two implementation milestones, with 50% deployment of the authorized constellation at the first milestone, requiring launch and operation within 6 years.[39] The implementation of national milestones should be regarded as a good effort to bring awareness on the satellite operators, but it is only complementary to the initiative of the ITU, which takes precedence in imposing the minimum requirements. Another modification in the FCC attention was related to "Spectrum Sharing among NGSO FSS Systems." The FCC decided that in the first instance and before resorting to a default mechanism, the non-GSO operators have to coordinate among themselves and discuss the technical operations in good faith with the aim of accommodating the systems.[40]

5.13   Access to Market: The Case of Africa

The success of the new business model for large constellations of satellites is not yet guaranteed. Challenges could be divided into (1) cost of the services and (2) market access. The company has to generate profit, but providing Internet in remote areas and expecting profit to support the business may not be reasonable. In addition to the Internet services, the operators may also target the GSM services, but the market access in those cases would be conditioned by licenses given by national telecommunication authorities. The mobile service operators are strongly positioned on the markets and have very competitive prices, making the market access a big rock to challenge including licenses.

Regarding prices, any prediction is challenging, especially when looking in the past nineties, when the constellation operators have failed, for example, Teledisc, Globalstar, Iridium, and Skybridge.[41] Regarding market access, OneWeb intends to

---

[37] Letter from William M. Wiltshire, Counsel to SpaceX, to Jose P. Albuquerque, Chief, Satellite Division (dated April 20, 2017).

[38] FCC 17-122, *Report and Order and further Notice of Proposed Rulemaking* [2017].

[39] Ibid. *supra* note 38, pp. 20–23.

[40] Ibid. *supra* note 38, pp. 15–18.

[41] Bender, R., *Launching and Operating Satellites: Legal Issues*, (pp. 54, Utrecht Studies in Air and Space Law - Vol. 18, The Hague, Boston: Martinus Nijhoff Publishers, 1998).

use the local capacity and work with local Internet service providers and mobile service providers as an extension of the existing networks and not as replacement.[42] This is important as the end user will be billed by the existing operators and not by OneWeb, which will only connect the devices on unlicensed frequencies or be using partner operator frequencies to provide a better coverage.[43] Then reusing the customer devices is important to reduce costs and coding in open source.[44] Next, the problem is about the visibility of the service; antennas are needed to capture such a service. For example, OneWeb has announced reaching an agreement valued more than $300 mil. with Hughes Network Systems for the production of its ground network system with gateways featuring multiple tracking antennas to support operations and traffic from and to LEO satellites.

A particular market that is appealing to large satellite constellations operators is Africa, and OneWeb announced to provide *fiber in the sky*.[45] The question is how to support the business model; for example, is smart agriculture and automatic technologies a priority in Africa?

During the 2018 World Summit on the Information Society (WSIS) session 189, *Megaconstellations offering new technology for an inclusive access to rural areas of the LDCs*, a possible partnership was mentioned between the World Meteorological Organization (WMO) and OneWeb and was presented as one of the solutions for concerns on financial support. The World Bank was also mentioned. The focus on least developed countries (LDCs) in Africa is a global concern, and political stability could attract funds and support the implementation of autonomous technology by making use of the systems that are already there. For large constellations, a pilot project could represent a first step for demonstrating the need and technology.[46]

The challenges that people face in Africa in regard to the proposed technology are adaptability and compatibility. In reference to adaptability, people are used to working with historic providers. In relation to compatibility, on the other hand, substantial efforts are necessary to provide access and connectivity.[47] Some constraints for the business model could be energy (electricity) because it is not available in remote areas. Also, transport is challenging, but the most important aspect raised by the African delegation was how to make the people stay in remote areas as the general trend is to move closer to the cities. Another concern by the local authorities could be in regard to money going out of Africa instead of attracting resources.[48] In this regard, at the ITU Telecom World 2018, South Africa's Minister of Communications, referring to the country's plans for a smart and inclusive digital

---

[42] Tony Azzarelly, WSIS2018, Geneva, Room Popov 2, session 189, 19 March 2018.

[43] See: http://www.oneweb.world/#rural-coverage-for-mobile-operators.

[44] Yvette Ramos, WSIS2018, Geneva, Room Popov 2, session 189, 19 March 2018.

[45] Ibid *supra* note 42.

[46] Betty Bonnardel, WSIS2018, Geneva, Room Popov 2, session 189, 19 March 2018.

[47] Senegal and Cote d'Ivoire, final discussions, WSIS2018, Room Popov 2, session 189, 19 March 2018.

[48] Ibid. *supra* note 47.

future, mentioned that the country aims at "participating not only as consumers but as owners of products" in the digital economy.[49]

Potential growth in Africa is around 80%, a reason why money made in information and communication technologies (ICTs) could finance development and information society in Africa.[50] The new technology should be embraced, and Africa should be aware of the trend of large constellations of satellites bridging the digital divide. Young people in Africa are better connected than any other generation before them, but even so the potential of growth is humongous and perceived by the leaders. For example, Kofi Annan as a leading figure for the UN and Africa advised to act locally but think globally, because of the interconnectedness benefits.[51] "The internet can bring learning to the most remote areas of the planet," he said.[52] The latest statistic referring to the proportion of household with access to the Internet places Africa in the last place with only 18%. On the opposite, Europe has the highest proportion with almost 85%.[53]

"Broadband infrastructure is now vital infrastructure, as essential as water and electricity networks," mentions the ITU in the 2018 Broadband Report.[54] The ITU talks about the fourth industrial revolution (4IR) and the transformation of societies due to significant technological advancements in computer science, data science, artificial intelligence (AI), and cloud computing, and also about the potential of 5G broadband to become disruptive.[55] But the driving force behind the "first-mover advantage" needs to include discussions of "last-mover disadvantage" and the need to strategically invest in the digital infrastructure. Least developed countries (LDCs) should not be left behind in the race to digitalize, the case of market access has to be discussed, and the Internet should be made available to the second half of the world's unconnected population, together with its benefits for education, health, transport, agriculture, and security in the new era. A short analysis of the market access policies will be presented in the next chapter.

---

[49] See: https://news.itu.int/south-africa-smart-digital-transformation/.

[50] Dr. Cissé Kane, President ACSIS-SCASI, WSIS2018, Room Popov 2, session 189, 19 March 2018.

[51] Kofi Annan, ITU interview May 15, 2013, YouTube.

[52] See: https://news.itu.int/technology-can-improve-the-state-of-the-world-kofi-annan/.

[53] ITU, Measuring the Information Society Report 2017, Volume 1, pp. 13–18, ISBN 978-92-61-24511-5 (Paper version), <https://www.itu.int/en/ITU-D/Statistics/Documents/publications/misr2017/MISR2017_Volume1.pdf>.

[54] Philippa Bigs (ed), "The State of Broadband 2018: Broadband Catalyzing Sustainable Development" (ITU, ISBN: 978-92-61-26431-4, pp. 6–10, September 2018) <https://www.itu.int/dms_pub/itu-s/opb/pol/S-POL-BROADBAND.19-2018-PDF-E.pdf> accessed: 23.10.2018.

[55] Kemal Husenovic, Iqbal Bedi and Sofie Maddens (ed), "Setting the Scene for 5G: Opportunities & Challenges", (ITU, ISBN: 978-92-61-27591-4, pp. 8–9; pp. 16–20, 2018) <https://www.itu.int/en/ITU-D/Documents/ITU_5G_REPORT-2018.pdf> accessed: 23.10.2018.

5.14   The Open Skies Policy as an Option for Market Entry

One may wonder what the open skies policy means for satellite telecommunications and whether it could give the large constellations of satellites operators a more facile access to foreign markets. "Open Skies" means that national regulators will not impose requirements or restrictions on the use of foreign satellite systems than on the use of domestic/national satellite systems.[56] In fact, Africa has in general an "open skies" policy for satellite market access, although Nigeria is questioning the policy and threatens to change that and to add licensing and fees, which the satellite industry always opposes.

As an example for supporting the "Open Skies" approach in Africa, OneWeb has contributed with comments to the public inquiry on the draft Guidelines on Commercial Satellite.[57] The reason for action is that under the provisions of Part 1, Article 4 (1), *General Terms and Conditions* of the draft Commercial Satellite Communications Guidelines released for public comment in February 2018 (Draft Guidelines), Nigeria requested licensing of the space segment, meaning that satellite service providers authorized by a foreign administration will be required to obtain from the Nigerian Communications Commission (NCC) an additional authorization. Also, in paragraph 16 (1) of Part 4, *Space Segment*, it was mentioned that the NCC will impose fees and charges for the space segment landing rights authorization of 10,000 USD for the life span of the satellite, necessary for making the satellite accessible to customers and end users in Nigeria.[58] OneWeb opposed the change in policy and expressed the view that Nigeria should continue to support international best practices, for instance to maintain open skies access to satellite services.[59] Upon OneWeb's request for clarification on the policy that no landing rights authorization should be required for non-Nigerian satellite operators, especially on a per-satellite basis, the NCC did not provide a clear answer but was neither against the benefits of the current "open skies" policy, mentioning only that it will review this provision on the basis of reciprocity and that "landing rights" will apply to a large number of satellites, if justified.[60] Because the 10,000 USD charge on a per-satellite basis is particularly relevant for the large constellations of satel-

---

[56] EMEA Satellite Operators Association (ESOA), "Open Skies Policy – Market Access Principles for Satellite Communications" (www.esoa.net; 2018).

[57] Via E-mail – March 12, 2018, OneWeb position to Mrs. Yetunde Akinloye, Head, Legal and Regulatory Services Department, Nigerian Communications Commission.

[58] Nigerian Communications Commission, "Commercial Satellite Communications Guidelines", (2017) <https://www.ncc.gov.ng/documents/743-guidelines-on-commercial-satellite-communication-2017-draft/file> accessed: 19.10.2018.

[59] *"Given local service providers are duly licensed by NCC to provide services to public, and they are fully compliant with all relevant regulatory requirements; Landing Rights shall not be required for satellite operators who intends to wholesale capacity to local licensed service providers."*

[60] Nigerian Communications Commission, "Report of the Public Inquiry on the Draft Guidelines on Commercial Satellite" (Article 6, p. 5, 30.04.2018) <https://www.ncc.gov.ng/documents/796-public-inquiry-on-the-draft-guidelines-on-commercial-satellite/file> accessed: 19.10.2018.

lites in LEO, which will be formed from hundreds to thousands of satellite systems, OneWeb has requested clarification on whether it is meant to address satellite systems as a whole and not on a per-satellite basis. The NCC only mentioned that the fee covers all the satellites owned by an operator; however, it will review its position on applicable fees.[61]

The request of the EMEA Satellite Operators Association (ESOA) for clarification related to the authorization of space segment satellite operators revealed the NCC policy for market access.[62] The NCC clearly stated in Article 15 on the Report of the Public Inquiry that a foreign space segment satellite operator that provides services through locally licensed service providers will not be required to get an authorization but must ensure that the local service provider is a licensee of the NCC.[63] This clarification is in line with the OneWeb policy mentioned in the chapter above and confirms its intention to work with local service providers without a direct contract with end users. Overall, ESOA's recommendations to NCC referred to the following: (1) maintaining the "Open Skies" benefits, (2) maintaining a level playing field for the authorization of satellite terminals, (3) clarifying that an authorization of space segment satellite operator is not requested for the foreign operator providing service through locally licensed service providers, (4) clarifying that the local presence requirement is not required for visiting Earth Stations in Motion (ESIMs) for a period of less than 6 weeks, (5) clarifying that the 10,000 USD fee is not a per-satellite charge but addresses the satellite system as whole, (6) clarifying that the type approval should be obtained by the antenna manufacturer and recognized by the NCC, and (7) assisting NCC to address the critical task of interference risks.

Currently, there are two different possibilities for market access: one is known as "open skies" policy, and the other, a more protectionist one, requires "landing rights." It is important to mention that these policies refer to access to the space segment. The "open skies" policy is particularly important if a foreign space segment satellite operator plans to provide services directly to end users and was regarded as a fundamental regulatory principle for liberalizing access to all national and foreign commercial satellites.[64] For instance, the open skies regulatory approach does not impose any requirements on market entry such as licensing fees or other formalities beyond registration of technical criteria on foreign satellite operators that ultimately allows nationally authorized service providers to choose any satellite operator or satellite service provider to distribute their specific services to areas required for

---

[61] Ibid *supra* note 60, Article 20, p. 9.

[62] EMEA Satellite Operators Association (ESOA), and Global VSAT Forum (GVF) "Comments on Draft Commercial Satellite Communications Guidelines" (12 March 2018): *"ESOA and GVF understand that this would only be the case if a foreign space segment satellite operator plans to provide services directly to end-users in Nigeria. Otherwise, in case services are provided through local licensed service providers, such authorization would not be required. We would appreciate a confirmation from the NCC on this point."*

[63] Ibid *supra* note 60, Article 15, p. 7.

[64] Ibid *supra* note 62, pp. 2–3.

their national or international end users. Basically, the "open skies" policy recognizes that the authorization of space stations from the satellite operators' administration (home licensing administration) is sufficient and there is no need to duplicate it in other countries.[65] In regard to market access recommendations, ESOA generally promotes the following open skies principles in order to reduce regulatory and market access barriers[66]:

- Make the provision of bare satellite capacity unrestricted.
- Provide national treatment for foreign satellite operators.
- Eliminate local entity and/or local presence requirements.
- Provide transparent and nondiscriminatory authorization procedures.
- Eliminate burdensome frequency coordination requirement.
- Eliminate monopolies.
- Allow transport of video signals and associated audio signals.
- Allow free circulation and use of satellite consumer terminals.
- Address security concerns adequately.

These principles are favorable policy principles to promote access market, including for the large constellations of satellites, and should not be considered as indivisible from the open skies policy and be analyzed separately; instead, it should be considered together with international best practices.

According to OneWeb, in general, half of Asia (and half of South America) do require "Landing Rights" (market access for the satellite operator), which the industry sees as protectionist to the domestic satellite industry and never an efficient way to get applications and products to consumers. Some countries (Russia, China, India, Indonesia) are particularly difficult, in that a purely "regulatory" application is insufficient. These countries demand "coordination" with their domestic operators before a "foreign" operator is allowed to offer services, and in practice, this means that those operators must be satisfied with the commercial situation, which in fact is not directly a real ITU coordination. So the foreign operator must work out a "commercial solution" with the domestic operators before it is allowed to offer satellite services in that country, which can take years. (Brazil also requires domestic coordination, but it really is just a spectrum coordination: it is not expected that the foreign operator work out some kind of "joint venture" with those operators.)

"The requirement of service licensing for satellite connectivity will only raise costs to end users," expressed OneWeb in a recent paper presented at the International Astronautical Congress 2018 in Bremen. The "Open Skies" policies for domestic licenses were presented as a benefit for end users and for operators, which with a simple registration could choose the best infrastructure, as opposed to "landing rights" or other domestic service licenses, which does not allow operators to do so.[67]

---

[65] Ibid *supra* note 62, p. 2.

[66] Ibid. *supra* note 56, pp. 4–6.

[67] Ruth Pritchard-Kelly, Yvon Henri, "To Fully Bridge the Digital Divide by 2027, Making Internet Access Available and Affordable for Everyone - the non-GSO Constellation Response (Regulatory Best Practices)" [2018] IAC-18, E7,2,12,x46476.

The emphasis was put on blanket licensing and exemptions from licensing, online application processes, and predictable timelines. A blanket authorization suggests no separate licensing requirements for each end-user terminal. The spectrum fees "per MHz" was also mentioned as a regulatory burden for the future services of large constellations of satellites, which will need to use large bandwidths. The European authorization regime was taken as an example in regard to licensing Earth Stations in Motion (ESIMs) of non-GSO systems. It was mentioned that the Electronic Communications Committee (ECC) adopted a series of decisions such as ECC/DEC/(17)04 in 2017, which exempted from licensing individual satellite terminals, and ECC/DEC/(18)05 in 2018, which exempted from individual licensing the ESIMs, both adopted using the OneWeb system on compatibility studies.

5.15   The Risk of Space Debris Posed by Large Constellations of Satellites in LEO

Concerns were raised about the potential of the large constellations of satellites to endanger future activities in outer space by negatively impacting the number of space debris population in LEO.[68] The reliability of critical systems and the functionality of small satellites, such as end-of-life disposal, are considered key issues for space debris mitigation. A high level of compliance in the disposal of these satellites in orbits with short residual lifetimes is desirable.[69]

At the opening of the European Centre for Space Law (ECSL) Practitioners Forum in April 2018, Dr. Johann-Dietrich Wörner, Director General (DG) of the European Space Agency (ESA), addressed large constellations of satellites mentioning that urgent improvements are needed in the reliability of small satellites, including the disposal function at the end of life compared with currently demonstrated capabilities. The ESA DG clearly underlined that currently 15% of the satellites forming the large constellations are space debris. From 1000 satellites, 150 are space debris, raising awareness about the risk for space debris posed by large constellations of satellites.

From the ITU's perspective, satellite orbits and management of radio spectrum could be negatively influenced by space debris, as it may cause potentially harmful interference for the activities of States and intergovernmental organizations. The role of the ITU for mitigation of space debris is currently addressed through ITU-R S.1003-2 for the environmental protection of the GSO. The ITU-R Recommendations

---

[68] Popova R., Schaus V., "The legal framework for Space Debris remediation as a tool for sustainability in outer space" [2018] <http://www.mdpi.com/2226-4310/5/2/55> accessed: 08.10.2018.

[69] Lewis H.G., Radtke J, Rossi A., Becl J, Oswald M., Anderson P., Bastida Virgili B, Krag H, "Sensitivity of the Space Debris Environment to Large Constellations and Small Satellites" [2018], <https://conference.sdo.esoc.esa.int/proceedings/sdc7/paper/507/SDC7-paper507.pdf> accessed: 20.10.2018.

are a set of international technical standards for management and use of the spectrum/orbit approved by the ITU Member States and ITU-R sector members.

5.16 The Role of the ITU in Mitigation of Orbital Collisions and Space Debris in Relation to Large Constellations of Satellites

Orbiting a large number of small satellites in significant constellations to provide broadband telecommunication services raises the ITU concerns about the protection of the frequency assignments and orbits registered in the MIFR.[70]

The ITU does not have yet an ITU-R recommendation for LEO. The ITU recommendation for GSO should be translated into a similar document that will focus on non-GSO satellite systems. The ITU recommendation will not be an alternative to IADC and UN guidelines but should be complementary in order to gain awareness for mitigation of space debris in LEO and contribute to the implementation of the guidelines in national legislation. The ITU-R can provide stricter guidelines, including the requirement to facilitate active debris removal. The ITU-R Recommendations are the highest nonbinding instrument of the ITU; their implementation is not mandatory, but they enjoy a high reputation and are implemented worldwide, *de facto* a standard. The private sector of the ITU is composed of, for instance, satellite operators, and the ITU can reach them directly.[71]

The Committee on the Peaceful Uses of Outer Space (COPUOS) issued in June 2018 the latest version of the *Guidelines for the long-term sustainability of outer space activities.* These guidelines mention the important role of the ITU in ensuring the equitable, rational, and efficient use of the radio spectrum and the various orbital regions used by satellites.[72]

It was emphasized the need for the ITU Member States to give particular attention to the long-term sustainability of space activities and facilitate the prompt resolution of identified harmful radio frequency interference.[73] It was also mentioned that spacecraft and launch vehicles that have terminated their operational phases should be removed from orbit in a controlled fashion or, if this is not possible, be disposed in orbits that avoid their long-term presence in LEO.

---

[70] Ibid. *supra* note 3, p. 241.

[71] ALLISON A. L., *The ITU and Managing Satellite Orbital and Spectrum Resources in the 21st Century*, (SpringerBriefs in Space Development. Cham: Springer International Publishing, 2014).

[72] Doc. A/AC.105/L.315, *Guidelines for the long-tern sustainability of outer space activities*, 61th session of COPUOS in Vienna, 20–29 June 2018.

[73] Ibid. *supra* note 72.

5.17 Conclusions

Securing frequency assignments for the new commercial activity in LEO represents a challenge. To answer the research question, the WP4A of the ITU made already important progress in addressing the need for an improved regulatory framework to address BIU frequency assignments specifically for non-GSO satellite networks. The goal is to avoid warehousing of spectrum and avoid administrations to block spectrum for others. The decision is expected to be taken at the WRC-19 based on the work of the WP4A in CPM for WRC-19 Agenda Item 7 Issue A—"Bringing into use of frequency assignments to all non-GSO satellite systems, and consideration of a milestone-based deployment approach for non-GSO satellite systems in specific bands and services."

Administrations do not have experience with large constellations of satellites, thus making the first milestone the most challenging. It is important for the WRC-19 to establish a clear framework to facilitate equitable access, rational use, and protection from harmful interference but also to accommodate the milestone deployment in case BIU happens before the WRC-19.

The "open skies" policy for market access is beneficial for the large constellations of satellites and means liberalizing access to all national and foreign commercial satellite service providers and operators without imposing excessive fees and charges on a per-satellite basis or the obligation to duplicate the space segment authorization given to the satellite operator by the home administration. The "open skies" policy is currently implemented in various degrees in a significant number of jurisdictions and is based on reciprocity. It is important for the regulators to consider the utility in their own region of international best practices in order to encourage the deployment of innovative satellite technologies and promote a low price for end users of the services such as those provided by non-GSO satellite systems, which are currently seeking market access.

This research also raised the problem of the risk of space debris posed by large constellations of satellites and the ITU involvement in LEO. The Member States' delegations at the ITU should propose a new instrument referring to large constellations of satellites in LEO at the ITU Council. Because the ITU is interested in protecting access to orbital position, assuring frequency allocation, and preventing any harmful interference, it should also be active in mitigating space debris in LEO. The ITU means of communicating with its members and tying relations with the industry could be regarded as an extremely relevant instrument for supporting space debris mitigation.

Claudiu Mihai Tăiatu graduated from the International Space University (ISU) Space Studies Program (SSP18) and from the Advanced LL.M. of Air and Space Law, IIASL, Leiden University, including a thesis on "Current Challenges and Legal Perspectives for Megaconstellations: Bringing into Use Frequency Assignments to a Space Station of a non-Geostationary Satellite Network." He is a prospective member of the International Institute of Space Law (IISL) and a member of the European Centre for Space Law (ECSL). He followed an internship in Geneva at the International Telecommunication Union (ITU) and a research scholarship in Rome at UNIDROIT. He was awarded with the 2017 IISL Diederiks-Verschoor Award for the paper "Space Traffic Management: Top Priority for Safety Operations." c.taiatu@yahoo.com.

Chapter 6
The Sustainability of Large Satellite Constellations: Challenges for Space Law

Vinicius Aloia

Abstract The number of satellites launched into outer space is increasing significantly. The quantity of space objects orbiting the Earth, especially in low Earth orbit, is causing serious concerns for space agencies as it poses a key threat to the sustainability of space activities. Many space agencies are attempting to reduce the generation of space debris by following non-legally binding space debris mitigation guidelines. These guidelines are common instruments for space law development in the 'post-treaty' era and are voluntary in nature. Consequently, no binding international norms regulating space debris exist today. Also, it has become apparent that these mitigation guidelines are not comprehensive enough and the current legal and regulatory framework does not adequately respond to the new challenges presented by large satellite constellations, particularly the proliferation of space debris. The aim of the author is to discuss some of the sustainability issues and legal challenges relating to future large satellite constellations. The goal is to present possible solutions and important proposals aimed at ensuring international cooperation as a key element in the sustainability of human activity in outer space, guaranteeing free access to outer space by all nations, continuing freedom of exploration and use and avoiding harmful interference.

6.1 Introduction: From Public to Private

Since the launching of Sputnik 1 on 4 October 1957, 'developments have occurred so rapidly that only those professionally concerned with developments in space can hope to keep track of them' (Jenks 1965).[1] However, the law has failed to keep up with technological development. This is of particular importance for the exploration and use of outer space.

Soviet Sputnik 1 was the first artificial satellite that was ever launched around the Earth. Sixty years ago, outer space was reserved for only the most powerful nations

---

[1] Jenks CW, *Space Law* (Stevens & Sons 1965) 3.

V. Aloia (✉)
University of Helsinki, Helsinki, Finland

© Springer Nature Switzerland AG 2019
A. Froehlich (ed.), *Legal Aspects Around Satellite Constellations*, Studies in Space Policy 19, https://doi.org/10.1007/978-3-030-06028-2_6

and the most dominant governments. There was a fast rise of technological progress of satellite applications in the following years after the launching of Sputnik, especially of communication and Earth observation. Nevertheless, satellites were predominantly of military interest only.

Private actors—or non-governmental entities[2]—were long excluded from space activities for many different reasons. Due to the obvious strategic and political importance of space, governments have historically hesitated to entrust anyone with their own military or governmental space agencies with the exploration and use of outer space. In addition, the immense development costs and too high financial as well as technological risks helped deter any potential inventors. The legal framework for private commercial activities in space was—and keeps being—rather insufficient, if not totally missing, even in countries where they were in principle welcome.

With time, space technology developed and, due to decreasing costs, it steadily began to attract private investors. Initially, predominantly public-private partnerships were put into effect and the possibility of using satellites for scientific research, for services of public interest and also for commercial undertakings, was recognised. Satellite applications were the first field of space activities to be opened up to private and commercial actors, most notably in the U.S.[3]

More recently, with regard to various aspects of space activities, there is a growing trend towards the privatisation and commercialisation of space activities, a phenomenon referred to as 'NewSpace',[4] even though the majority of outer space activities are still carried out by governmental entities.[5] In fact, approximately 80% of all space activities are still being financed with public money.[6] There is an undeniable and clear democratisation of space most notably due to the profitability of space activities, but nevertheless this new trend has to be taken with a grain of salt.

'Usually, the privatisation of an area goes hand in hand with its commercialisation' (Edith Walter 2011),[7] thus with making money and profits. Different actors take their chances, competition arises between them and, eventually, a new market emerges. Space commercialisation, in particular, can be defined as 'the use of equipment sent into or through outer space to provide goods or services of commercial value'.[8] And space is a very lucrative place.[9]

---

[2] See Article VI of the Treaty on Principles Governing the Activities of States in the Exploration and Use of Outer Space, including the Moon and Other Celestial Bodies (OST).

[3] Brünner C and Soucek A, Outer Space in Society, Politics and Law (Springer 2011) 494.

[4] Hobe S and others, Cologne Commentary on Space Law (Carl Heymanns Verlag 2009) 14.

[5] Ibid 14.

[6] https://www.space.com/2401-space-access-private-investment-public-funding-debate.html.

[7] Ibid 493.

[8] Ibid 493.

[9] Brünner and Soucek (n 4) 135.

Now, there are new initiatives to foster space commercialisation.[10] The commercial space industry, using multimillion-dollar satellites and rockets, is increasingly playing a more prominent part in our everyday lives[11] and, in the process, growing the interdependence between what is happening hundreds of kilometres up in space and down below on Earth. Although this may seem like a new phenomenon, in reality, profit-seeking space enterprise has been going on for decades. However, this intense commercialisation of space activities and this shift from public to private comes at a potentially irreparable cost. Let's see.

6.2   Communication Satellites: From GSO Behemoths to LEO Small Cubes

There are various kinds of satellites used for commercial purposes. It is important to highlight that satellites are dual-use applications, serving both military and civilian purposes.[12] [They] can be divided into, i.a., earth observation, meteorological, communication and navigation satellites depending on their payload and practical use.[13] These satellites not only differ in function, size and design but also vary in particular orbits that they are placed in.

Communication satellites, for instance, are mostly placed into geosynchronous orbit (GSO), in which they travel in a circular path around Earth in the plane of the equator, at an altitude of c. 35,786 km.[14] By placing a satellite in this position, it appears from any given point on Earth to be stationary. This is possible because a satellite in GSO orbits at a speed of 3.1 km/s (11,600 km/h) and completes a full rotation around the planet roughly every 24 h, in accordance with the Earth. Placed at this position, a constellation of just three equidistant communication satellites in the same orbit is enough to ensure full coverage of the Earth.[15]

Note, satellite-based communication was the first commercial space application and has been used particularly in areas such as telephone transmission; video

---

[10] Martin, Gary (25 January 2016). "NewSpace: The "Emerging" Commercial Space Industry" (PDF). *nasa.gov*. NASA. Archived (PDF) from the original on 24 June 2016. Retrieved 16 September 2018.

[11] https://www.extremetech.com/extreme/268062-5-reasons-space-exploration-is-more-important-than-ever.

[12] Objects serving both military and civilian purposes. *See* Lyall F and Larsen PB, *Space Law a Treatise* (Ashgate 2009) 403.

[13] Hobe and others (n 5) 8.

[14] Tronchetti F and Dunk FG von der, *Handbook of Space Law* (Edward Elgar 2015) 798.
   See also: For a description of the characteristics of the geostationary orbit see: K.U. Schrogl, Questions Relating to the Character and Utilization of the Geostationary Orbit, in *International Space Law in the Making: Current Issues in the United Nations Committee on the Peaceful Uses of Outer Space* (Eds. K.U. Schrogl & M. Benkö) (1993); S. Cahill, Give Me My Space: Implications for Permitting National Appropriation of the Geostationary Orbit, 19 *Wisconsin International Law Journal* (2001), 231; F. Lyall, *Law and Space Telecommunications* (1989), 388.

[15] Ibid.

conferencing; television and radio industry; mobile communication to aircraft, ships and mobile phones; telemedicine; and the computer industry.

Although not legally defined, a satellite constellation consists of a space segment of two or more artificial satellites working with a coordinated ground coverage, operating together under shared control, and synchronised so that they overlap well in coverage, independently of their orbit.[16]

After 60 years, the geostationary orbit position is crowded. New ideas of using the space around the Earth to maximise profits and expand global coverage are being developed. These include the launching of large satellite constellations (also known as mega-constellations) made up of a vast network of small satellites. Before NewSpace and the commercialisation of outer space, satellite applications were often very costly, required an 'army' of scientists and engineers and a huge amount of money in investment (usually millions of dollars), on top of decades of development and testing. Today, thanks to technological advancements, especially relating to computing power,[17] satellites and space applications can literally be developed by anyone.[18]

Large satellite constellations are yet to be launched non-GSO fixed-satellite service system (FSS) infrastructures in low Earth orbit (or, in some cases, medium

---

[16] https://en.wikipedia.org/wiki/Satellite_constellation.

[17] See. Franklin D, Megatech: Technology in 2050 (The Economist 2018).

Everyone knows that modern computers are better than old ones. But it is hard to convey just how much faster, for no other consumer technology has improved at anything approaching a similar pace. The standard analogy is with cars: if the car from 1971 had improved at the same rate as computer chips, then by 2015 new models would have top speeds of about 420 million mph. That is roughly 2/3 the speed of light, or fast enough to drive around the world in less than a fifth of a second. If that is still too slow, then before the end of 2017 faster-than-light models that can go twice as fast again will begin arriving in the showrooms.
[…]
This blistering progress is a consequence of an observation first made in 1965 by one of Intel's founders, Gordon Moore. Moore noted that the number of components that could be crammed onto an integrated circuit was doubling every year. Later amended every two years, "Moore's law" has become a self-fulfilling prophecy that sets the pace for the entire computing industry. Each year firms such as Intel and the Taiwan Semiconductor Manufacturing Company spend billions of dollars figuring out how to keep shrinking the components that go into computer chips. Along the way, Moore's law has helped to build a world in which chips are built in to everything from kettles to cars (which can, increasingly, drive themselves), where millions of people relax in virtual worlds, financial markets are played by algorithms and pundits worry that AI will soon take all the jobs.
[…]
Moore's law has made computers smaller, transforming them from room-filling behemoths to svelte, pocket-filling slabs. It has also made them more frugal: a smartphone that packs more computing power than was available to entire nations in 1971 can last a day or more on a single battery charge. But the most famous effect has been to make computers faster. By 2050, when Moore's law will be ancient history, engineers will have to make use of a string of other tricks if they are to keep computers getting faster.

[18] https://www.foreignaffairs.com/articles/space/2015-04-20/democratization-space.

Earth orbit (MEO)), for telecommunication purposes. There is a wide variation on non-GSO FSS satellite on an LEO with a single inclination, e.g., satellite systems consisting of thousands of satellites on LEO circular orbits with a different inclination and satellite systems consisting of thousands of satellites in more than hundreds or thousands of orbital planes.[19] Furthermore, they require mass manufacturing of satellites with low cost per unit, frequent and flexible multi-payload launches, a population of several orbital planes—which is already very challenging for two or three satellites—and operational reliability of the satellites. Traditionally, telecommunications, broadcasting and meteorological satellites are launched into geostationary orbit (GSO).[20]

Unsurprisingly, there is a considerable amount of legal questions relating particularly to large satellite constellations. For the sake of this short dissertation, however, I focus on the technical, legal and political challenges regarding the sustainability of such endeavours, i.e., the environmental implications of having such a vast number of space objects in orbit around the Earth, and how non-legally binding space debris mitigation adopted by international organisations and nongovernmental entities—to ensure the sustainability of space activities in the future—are tackling the issue.

As satellite constellations grow larger—a phenomenon promoted by ambitious programmes of aerospace companies like SpaceX (StarLink) and OneWeb—governments and international organisations are beginning to get seriously worried about orbital debris.[21]

6.3 The Space Industry: Democratisation of Space and Standardisation as the Key

Standardisation of space applications for mass manufacturing purposes, for instance, is one of the key aspects that allow for the lowering of costs when developing new satellites.

A standard CubeSat spacecraft[22] is a type of *miniaturised* satellite for space research that is made up of multiples of $10 \times 10 \times 10$ cm cubic units. CubeSats are considerably lighter than regular satellites and often use commercial off-the-shelf (COTS) components for their electronics and structure. To be financially viable,

---

[19] Matas, Attila, *ITU Space Regulation - A Key Element to Access Space,* IAC-18.E7.7-B3.8.6.x42748.

[20] Lyall F and Larsen PB, Space Law A Treatise 2nd Edition (Taylor and Francis 2017) 248.

[21] https://www.theverge.com/platform/amp/2018/9/28/17906158/nasa-spacex-onewebsatellite-large-constellations-orbital-debris.

[22] See Terms and definitions of the IADC—Spacecraft—*an orbiting object designed to perform a specific function or mission (e.g. communications, navigation or Earth observation). A spacecraft that can no longer fulfil its intended mission is considered non-functional. (Spacecraft in reserve or standby modes awaiting possible reactivation are considered functional.)*

large satellite constellations will require the mass production of satellites, and such characteristics are of significant importance.

As of today, CubeSats are commonly launched as secondary payloads on a launch vehicle. Piggybacking on larger commercial launches contributes to the affordability of the mission; unlike full-sized spacecraft, CubeSats have the ability to be delivered into space as cargo and then put in orbit by deployers on the International Space Station (ISS). This presents an alternative method of achieving orbit apart from launch and deployment by a launch vehicle.[23] Over 800 CubeSats have been launched as of April 2018.

CubeSats have such a significance that some of them became the first national satellites of their respective countries, being launched by universities,[24] states or private companies. The first Finnish satellite, Aalto-2,[25] was developed by a team of researchers and students from Aalto University in Finland.[26] The Nanosatellite and CubeSat Database lists over 2000 CubeSats that have been and are planned to be launched since 1998.[27]

Although the GSO has long been the orbit of particular concern when talking about telecommunication satellites (and demand is still greatest for orbital slots in GSO[28]), attention must be focused on communication satellites (especially for broadband services) that are placed in the close vicinity of the Earth as part of LEO, generally below 2000 km, systems that the US Federal Communications Commission, in particular, has been licensing since 1994.[29] LEO is the region of greatest interest for space activities today and in the near future.[30]

In spite of its advantages for telecommunication satellites, GSO has its problems: due to its distance, it is relatively hard to get to the GSO, c. 35,786 km above the Earth's surface. In addition, in order to reach its target, the radio signal needs to be significantly powerful. This translates into large and robust satellites, expensive to be developed and manufactured. Satellites in LEO tend to be smaller than satellites in GSO. Also, radio signals move at best at the speed of light. This speed limitation means that the communication is commonly delayed (usually 1 s each way). It takes time for the signal to reach the satellite from the ground station and then from the satellite to the final recipient.

---

[23] "In-Space Satellite Construction May Be Coming Soon". *Space.com*. Retrieved 2018-09-29.

[24] https://www.nasa.gov/directorates/heo/home/CubeSats_initiative.

[25] Aalto-2 was the first Finnish satellite launched into space, but not the first developed Finnish satellite.

[26] https://yle.fi/uutiset/osasto/news/aalto-2_finlands_first_satellite_launched_into_orbit/9571387.

[27] Kulu, Erik. "Nanosatellite & CubeSat Database". *Nanosatellite & CubeSat Database*. Retrieved 28 September 2018.

[28] The population of space objects in GEO is also the most diverse as regards ownership: over 35 countries and organisations own assets there. Cosmic Study on Space Traffic Management 2006, 25.

[29] Viikari L, The Environmental Element in Space Law: Assessing the Present and Charting the Future (Nijhoff 2008).

[30] Cosmic Study on Space Traffic Management 2006, pp. 24, 26. In 2003, approximately 43% of all space missions to reach Earth orbit were LEO missions, while GEO missions accounted for 38%. Ibid.

Although the placement of these satellites into LEO requires far more satellites to ensure complete global coverage, there is still the advantage of much short propagation delays. In addition, the geostationary orbit is considered a limited natural resource, and the slots are physically limited. Article 44.2 (196) of the 2006 version of the ITU Constitution states that Member States shall bear in mind that 'any associated orbits, including the geostationary orbit, are limited natural resources and that they are to be used rationally, efficiently and economically, in conformity with the Radio Regulations, so that countries or groups of countries may have equitable access to those orbits (…) taking into account the special needs of the developing countries and the geographical situation of particular countries'.[31]

Large satellite constellations can solve this problem with hundreds or even thousands of small satellites in lower orbits. To illustrate the increasing interest of designers and operators on large constellations in the LEO region since the mid-1990s, the pioneering Iridium constellation must be mentioned with 66 active and 6 spare satellites and Orb-Comm constellation with (nominally) 48 satellites. Other companies consider LEO constellations that are orders of magnitude to be larger, consisting of tens, hundreds or even thousands of satellites, orbiting at about the same altitude and orbital plane, with the aim of providing continuous global communication[32] and Internet services with high data rates,[33] plastering the whole planet and providing global coverage.

Some prominent, potential competitors for this market segment are, i.a., SpaceX, with a constellation of c. 4000 satellites at 1100 km altitude; Boeing, with a constellation of 2960 satellites at 1200 km altitude; and OneWeb, with a constellation of c. 720 satellites at 1200 km altitude. Earlier this year, OneWeb has filed before the FCC for authorisation to launch a further 1200 more satellites.[34] SpaceX Starlink has the objective of eventually building a low-cost, satellite-based broadband network capable of delivering Internet access to the entire globe. According to the United Nations Office for Outer Space Affairs (UNOOSA),[35] there are currently roughly 1500 satellites in orbit around Earth, along with some 2600 inactive satellites. SpaceX will need to launch over 4000 satellites into orbit to achieve its desired global coverage.[36]

As can be seen, most of these constellations will be released at an altitude of approximately 1000 km and use low-thrust manoeuvres to reach their assigned orbits. To guarantee a continuous service, large satellite constellations will inevitably need to regularly replace ageing satellites through new launches.

---

[31] Lyall F and Larsen PB (n 21) 251.

[32] van der Ha, J. C. (ed.) Mission Design and Implementation of Satellite Constellations (Kluwer, Dordrecht, 1998).

[33] https://spacenews.com/us-regulators-approve-spacex-constellation-but-deny-waiver-for-easier-deployment-deadline/.

[34] https://spacenews.com/oneweb-asks-fcc-to-authorize-1200-more-satellites/.

[35] https://www.pixalytics.com/sats-orbiting-the-earth-2018/.

[36] https://www.digitaltrends.com/cool-tech/spacex-starlink-elon-musk-news/.

6.4   Dilemmas of NewSpace: The Problem of Space Debris

Many satellites are perfectly operational when they run out of fuel, which is crucial for keeping the satellite in its orbit. Although all instruments may still be intact onboard a space object, when operators can no longer manoeuvre them, they decommission the spacecraft. This is unsustainable in the long term. Space operators have started planning and working on the possibility of refuelling, repairing and even resurrecting satellites in outer space. This is called 'on-orbit servicing', which is another possibility to militate the accumulation and clogging of orbital debris.[37]

Additionally, to be environmentally compatible, the decommissioned satellites would need to perform end-of-life disposal activities that are in accordance with international debris mitigation guidelines (e.g., with remaining orbital lifetimes of less than 25 years). However, when considering large satellite constellations of thousands of satellites, a lifetime of 25 years does not seem adequate anymore. Also, according to the mitigation guidelines, a failure rate of 10%, in general, was found to be acceptable under current space traffic conditions. This is certainly not sustainable when the LEO launch rates increase drastically.[38]

Remarkably, the Iridium system is already operational. Although some measures have been taken to minimise the risk of impact, the possibility of disruptive collisions simply cannot be ruled out, as the selected altitude corresponds to a peak of the expected accumulation of debris. For a reasonably sized constellation, for instance, the risk of a catastrophic impact is about 10% per decade. The US government registered over 300,000 potential debris collisions in 2017 alone—and the problem is getting much worse.[39]

The architecture of a given constellation in LEO will make a break-up event particularly dangerous and vulnerable to the consequences of an impact, which may set up a kind of chain reaction, triggering more collisions, an event called Kessler syndrome. The Kessler syndrome, as defined by Professor Don Kessler of NASA in 1978, predicts that these sorts of impacts will become more and more common and create tinier non-trackable bits of debris to clog up space around the Earth. According to the theory, large objects will smash into other large objects and break them into smaller ones. With so much junk for these smaller bits to hit, a chain reaction could be set off that ends up with billions of tiny pieces way too small to be tracked, making it impossible to leave Earth without encountering an impact.[40] This is of particular significance when debris orbits lead the fragments to encounter satellites revolving around the Earth in the opposite direction; this makes head-on collisions possible, with higher impact speeds and greater collision probability.

---

[37] Rao RV and Abhijeet K, "Recent Developments in Space Law" [2017] 112.

[38] http://adsabs.harvard.edu/abs/2016cosp...41E.153B.

[39] https://www.businessinsider.com/space-junk-collision-statistics-government-tracking-2017-2018-4?r=US&IR=T&IR=T.

[40] https://wiki2.org/en/Kessler_syndrome.

What is more, the smaller the satellites' size, the more likely they are to be non-manoeuvrable space objects, making them more vulnerable to be hit by pieces of debris. The fact that there are many possible different orbits and orbital planes for LEO satellites further adds to challenges in coordination. On the other hand, objects in LEO are much closer to Earth and have more unstable orbits than those in GSO,[41] the result being shorter hazardous lifetimes for debris. The environmental consequences of a collision in LEO may be less severe, but the much higher relevant velocities of objects there mean that the risks posed by debris are nevertheless far greater.

Currently, there are an estimated 1 980 active satellites in orbit.[42] The problem is not exclusive to Earth. In the future, if human activities in and around the Moon increase significantly, lunar orbits may also become congested with man-made debris. There, the situation could deteriorate much faster because the Moon has no appreciable atmosphere that could burn up debris.

With the perspective of a possible increase of constellations of the intact LEO objects by almost 100%, and a LEO mass increase by up to 90%, members of the Inter-Agency Space Debris Coordination Committee (IADC), among them ESA, have performed studies on the effect of large LEO constellation deployments on the long-term space debris environment evolution.[43]

These analyses indicated that a major criterion for maintaining an acceptable environmental footprint for a large LEO constellation is the reliability of its post-mission disposal of decommissioned satellites in accordance with existing debris mitigation guidelines. This reliability should exceed 90%. However, current compliance levels for LEO spacecraft that require an induced end-of-life de-orbit have not reached 20% in any of the past 25 years. To reach the 90% disposal reliability is hence a major technical challenge.

The increasing quantity of satellites orbiting the Earth is causing concern for space agencies. A 1 cm fleck paint around the size of a fingernail travelling at a typical collision speed of 35,000 km/h has the kinetic impact energy of an object weighing 250 kg at 96 km/h. A 10 cm object has the impact energy of around 7 kg of TNT. Several space shuttle windows needed replacing after being hit by what was later identified to be flecks of paint.[44]

A solar array removed from the Hubble Space Telescope in 1993 showed hundreds of tiny impact holes. A piece of micro debris cut a hole in the solar panel of a Copernicus Sentinel 1A satellite, and in May 2017, astronaut Tim Peake revealed how a paint flack, believed to be just a few thousandths of a millimetre in diameter made a serious dent in the Coppola window of the ISS.

---

[41] According to the IADC Space Debris Mitigation Guidelines (*see more* below), is an "Earth orbit having zero inclination and zero eccentricity, whose orbital period is equal to the Earth's sidereal period".

[42] https://www.ucsusa.org.

[43] Klinkard H, *Large satellite constellations and related challenges for space debris mitigation.* The Journal of Space Safety Engineering (Elsevier).

[44] http://curious-droid.com/254/will-space-junk-end-modern-way-life-kessler-syndrome/.

The ISS can be moved out of position if an impact is predicted, which is not the case for most small satellites. Impacts have a significant chance of occurring, especially for large structures that remain in orbit for a long time. Particularly at risk are satellites at altitudes of about 800 and 1400 km, where there is already a high density of orbiting bodies and atmospheric drag is not effective in removing small fragments.

The Space Surveillance Network is currently tracking more than 750,000 pieces of debris of which 16,000 are 10 cm in diameter or larger, while most of them are between 1 and 10 cm.[45] There are thought to be more than 150 million pieces smaller than a 1 cm. Over c. 8500 vehicle flights have been launched since 1957.[46] Less than half of the satellites contained on those missions are currently in orbit, around 3000, and even fewer still are operational—1738 according to the Union of Concerned Scientists.[47]

Debris may not constitute 'harmful interference' as such, but it is a potential menace.[48] That is because the ITU registers physical GSO slots, as well as corresponding radio frequencies for different services by geographical region. It also, i.a., 'coordinate[s] efforts to eliminate harmful interference and to improve the use made of the radio-frequency spectrum for radio-communication services and of the geostationary-satellite and other satellite orbits'.[49] There is no exact regulatory definition in the Radio Regulations of the ITU related to non-GSO fixed-satellite service (FSS). A non-GSO FSS system can be considered as 'a constellation comprised of a group of satellites, operating in the frequency bands allocated to the Fixed-Satellite Service (No.1.21), with similar characteristics and functions, operating in similar or complementary orbital planes under a shared control for a coordinated ground coverage'. Non-GSO FSS systems are intended to 'provide high-speed, low latency broadband services, including Internet connectivity, throughout the world, including locations which cannot be reached using GSO satellites'.[50]

At the larger end of the scale of the debris includes defunct satellites, empty rockets staged, discarded shields and even lens caps. The oldest known item still in orbit is NASA's Vanguard 1 satellite and its upper stage launch rockets, which were launched in 1958. Actually, in spite of the impressive numbers, defunct satellites only make up a fraction of all orbital debris.

Even flecks of paint can pose serious risks to missions in Earth orbit. In 2002, an unidentified item was found to be a burnt-out stage from the Apollo 12 Moon Mission launched in 1969. Flecks of paint become detached from the dead vehicles as a result of the extreme ultraviolet radiation. Although they are small, they orbit at extremely fast speeds, and that means they can pack a serious punch if they hit anything.

---

[45] https://swfound.org/media/99971/wright-space-debris_situation.pdf.

[46] http://claudelafleur.qc.ca/Spacecrafts-index.html.

[47] Union of Concerned Scientists, 'UCS Satellite Database' www.ucussa.org/nuclear-weapons/space-weapons/satellite-database#.Wh3_0kqnGM8.

[48] Lyall F and Larsen PB (n 21).

[49] Viikari L (n 29) 87.

[50] Matas A (n 20).

NASA and other space agencies know that pieces of debris moving up to unfathomable speeds pose a serious threat to vehicles and anyone aboard. Nevertheless, surprisingly enough, considering the gravity of the situation, there have actually been very few large-scale collisions to date.

In 2007, however, China tested an anti-satellite (ASAT) weapon by destroying a defunct probe with a rocket in a relatively high orbit.[51] That explosion created the largest break-up in space history, adding a further 3000 new tractable piece of space debris around the planet. Two years later, in 2009, a dead Russian probe, Cosmos, collided and destroyed an operating US commercial communication satellite Iridium. The Aerospace Corporation estimated that the collision created a further 200,000 more tiny bits of debris.[52]

The expected lifetimes for debris depend primarily on its location: in low Earth orbits, the air drag of the upper reaches of the atmosphere will eventually cause the debris to decelerate and heat up so that it breaks up under friction, whereas in higher orbits the atmospheric drag is virtually nonexistent. Despite the cleaning effect of the atmospheric drag, it has been calculated that if removal of LEO spacecraft at the end of their life is not conducted within 25 years or so, we can expect a marked increase in the number of accidental collisions later in this century.[53]

The Earth has the capacity to self-clean itself. Although, unfortunately, this does not last for long, as all the objects are quickly replaced by objects descending from higher orbits. The problem is that we are launching objects faster than the Earth can manage to clean up out the rubbish. So what is being done?

6.5   Concepts of Space Law

6.5.1   Article VI of the Outer Space Treaty: Authorisation and Supervision

Undeniably, the NewSpace Economy[54] is reshaping the market. Private companies carrying on space activities must be competitive enough to generate enough profit to justify their activities. Regarding non-governmental entities, the most important

---

[51] https://www.theguardian.com/science/2007/jan/23/spaceexploration.china. See also: On 11 January 2007, the People's Liberation Army of China conducted its first anti-satellite (ASAT) weapons test, destroying, with a ballistic missile, one of its own weather satellites in space, at about 530 miles in LEO. The explosion created a debris cloud of thousands of metal particles creating collision risks for some 700 spacecraft orbiting in LEO.

[52] Ted Muelhaupt, *The Collision of Iridium 33 and Cosmos 2251* (Aerospace Corporation Crosslink Magazine, December 2015) http://www.aeropsace.org/crosslinkmag/fall-2015/the-collision-of-iridium-33-and-cosmos-2251/.

[53] Viikari L (n 29) 36. *See* "Accidental Collisions of Cataloged Satellites Identified" 2005, p. 2.

[54] https://www.esa.int/About_Us/Ministerial_Council_2016/What_is_space_4.0—*Commercialisation and privatisation of outer space* https://www.bloomberg.com/news/features/2018-07-26/welcome-to-the-new-space-age. See also: https://www.businessfinland.fi/en/for-finnish-customers/services/build-your-network/digitalization/new-space-economy/.

regulation in the space treaties is Article VI of the Outer Space Treaty (OST) which stipulates that the "appropriate State" is responsible for all activities of non-governmental entities under its jurisdiction, especially for their authorisation and continuing supervision. Also, Article VI dictates that states shall bear international responsibility for the conformity with international space law of national activities in outer space, also if carried on by non-governmental entities. 'Responsibility shall be borne especially for assuring that such activities are carried out in conformity with the other provisions set forth in the Outer Space Treaty'.[55]

However, further detailed regulation of those private activities is left to later international agreements or even national law.[56] What is clear, nevertheless, is the need to implement the provisions of the treaties. A licensing requirement—a prerequisite for authorisation—ensures compliance with international law by the licensee, since the State is responsible for the activities purported by private entities, including the obligation to compensate a State for international liability claims. In view of authorisation and supervision requirements, monitoring mechanisms are also important.

With increasing privatisation and commercialisation of space, States will ever-more draft national space law. Unfortunately, however, the result is a wide variety of legislation—fragmentation of international space law—even as to fundamentals such as scope, which raises questions on the need for harmonisation.

6.5.2 Non-legally Binding Instruments: Is It Too Late for Space Debris Mitigation Guidelines?

Attention to the problem of space debris has significantly grown in recent years. Although there is still no agreed binding definition as to what a 'space debris' is,[57] this term has been increasingly used in deliberation within UNCOPUOS. A Technical Report on Space Debris (1999) by the Scientific and Technical Subcommittee of the UNCOPUOS uses the following definition: 'Space debris are all man-made objects, including their fragments and parts, whether their owners can be identified or not, in Earth orbit or re-entering the dense layers of the atmosphere that are non-functional with no reasonable expectation of their being able to assume or resume their intended functions or any other functions for which they are or can be authorised.'[58]

The UN space treaties do not specifically address space debris, mostly because it was not seen as an issue at the time the treaties were negotiated. International dis-

---

[55] Hobe and others (n 5) 104.

[56] Brünner and Soucek (n 4) 504, *See* Treaty on Principles governing the Activities of States in the Exploration and Use of Outer Space, including the Moon and Other Celestial Bodies.

[57] Ibid 32.

[58] Ibid. See also: The term 'debris' could possibly also be replaced by 'pollutant', 'contaminant', or 'flotsam', for instance. *See* Fasan 1993, p. 282. Even space 'garbage' and 'wreckage' have been used. *See* Jasentuliyana 1998, p. 139 and Góbriel 1987, p. 113, respectively. Para. 6.

cussions on the regulation of space debris started only in the early 1980s, but it was only in 1993, upon the initiative of the world's major space agencies, that the Inter-Agency Space Debris Coordination Committee IADC was established. The thinking is that the problem for micro and small debris cannot be solved. Therefore, organisations must work to prevent the satellites and other space objects from colliding with each other, avoiding debris in the first place.

These guidelines, which became the basis for the space debris mitigation guidelines developed and adopted by UNCOPUOS in 2009, establish a series of measures and good practices aimed at reducing the risk of creation of debris.

These guidelines are voluntary in nature and not legally binding under international law. Consequently, no binding international norms regulating space debris exist today. In spite of the problem of enforceability of the instruments, space agencies have been implementing the guidelines for over a decade. The guidelines encompass the following seven provisions: limit debris released during normal operations, minimise the potential for break-ups during operational phases, limit the probability of accidental collision in orbit, avoid intentional destruction and other harmful activities, minimise potential for post-mission break-ups resulting from stored energy, limit the long-term presence of spacecraft and launch vehicle orbital stages in the LEO region after the end of their mission, and limit the long-term interference of spacecraft and launch vehicle orbital stages with the GSO region after the end of their mission.[59]

Furthermore, several states have included in their national space legislation provisions on space debris mitigation and prevention.[60] These provisions are obligatory for the actors, both of governmental and non-governmental nature, which have been authorised by those states to carry out activities in outer space. The insertion of these types of provisions in national space laws can be seen as an instrument to transform non-obligatory international norms into rules that are enforceable at least on a national scale.[61] However, the employment of normative reference to such instruments of non-legally binding character in domestic space legislation raises questions of theoretical and practical character. Compliance with 'internationally recognized standards and guidelines', for instance, is a frequent condition in domestic space legislation for the authorisation of non-governmental space activities in accordance with Article VI of the Outer Space Treaty. This legislative practice does not necessarily comply with domestic legislative requirements of being clear, specific and unequivocal.

Furthermore, future satellite large satellite constellations are expected to have a maximum orbital lifetime of 25 years, after which either they are de-orbited into the atmosphere to burn up or they have to be moved into a higher orbit commonly referred to as the graveyard orbit.[62] Also, the transfer to a graveyard orbit should be

---

[59] Brünner and Soucek (n 4) 605.

[60] See, for example, The Finnish Act on Space Activities (63/2018).

[61] Viikari L (n 29) 88.

[62] IADC Guidelines, *supra* n. 164. Re-orbiting of LEO spacecraft into orbits where the residual lifetime of debris is 25 years at most is an efficient measure for stopping the accumulation of space

carried out with particular caution in order to avoid radio frequency interference with active satellites.

Another instrument is the ISO Standard 24113 'Space Systems – Space Debris Mitigation Requirements', 2011: the international standard establishing the design and operation requirements to minimise the impact of space operations on the orbital environment.[63]

All of these documents, however, were conceived without having in mind large satellite constellations. In spite of this legislative effort, it has become apparent that these mitigation guidelines are not comprehensive enough and the current legal and regulatory framework does not adequately respond to the new challenges presented by NewSpace and large satellite constellations. In addition, the guidelines are non-legally binding instruments under international law, as mentioned before, and there is still a considerable amount of countries that have not yet signed and/or ratified some or any of the UN treaties in the first place.

6.5.3   Technical Solutions: A Way to Go?

NASA proposed a solution to the problem called Space Debris Elimination (SpaDE).[64] This option was necessary because there was simply no practical method of debris removal to date. SpaDE would shoot pulses of atmospheric gas at targets that would destabilise their orbits and force them to re-enter the Earth's atmosphere and burn up.

Meanwhile, ESA is working on its own solution. Also agreeing that the problem of micro and small debris cannot be solved, they are also going after the larger pieces. ESA is working on an application called the e-DeOrbit, which would use a net to a robotic arm to catch a large piece of junk and draw them towards the atmosphere where both the hunter machine and the piece of junk would burn up.[65] The first launch of e-DeOrbit is expected to happen in 2023.

Another interesting initiative, the British RemoveDEBRIS satellite, has become the first ever to clear up some space junk, using its onboard net technology.[66] The satellite was built by a consortium of space companies and research institutions led by the Surrey Space Centre.

---

debris. This option is not economically feasible for higher-altitude orbits, however. For GEO space- craft, the best end-of-life disposal mechanism currently is re-orbiting to a graveyard orbit at least 235 km above the nominal GEO altitude. At the moment, only half of all satellites are properly re-orbited. *See also* Tronchetti F and Dunk FG von der, *Handbook of Space Law* (Edward Elgar 2015) 750.

[63] Soucek A and others, *Space Law Essentials* (NWV Verlag GmbH 2016).

[64] https://www.nasa.gov/offices/oct/early_stage_innovation/niac/gregory_space_debris_elimination.html.

[65] http://www.esa.int/Our_Activities/Space_Engineering_Technology/Clean_Space/e.Deorbit.

[66] https://www.forbes.com/sites/bridaineparnell/2018/09/19/removedebris-uk-satellite-is-the-first-to-clean-up-space-junk/#3f07918d5940.

Human ingenuity suggests that we would be able to solve a problem if we had to, but it makes much more sense to find the solution before the Kessler syndrome[67] occurs. How long we have got to do this is anybody's guess. In the meantime, with more and more nations joining the space race and more than a hundred vehicle launching every year—and that figure is set to increase—the problem is only going to get worse.

6.5.4 Innovative Ideas for Challenging Problems: Space Traffic Management

The Outer Space Treaty was intended to guarantee free access to space and to foster international cooperation.[68] On the 52nd anniversary of the OST, is space law and regulation meeting the NewSpace paradigm? The freedom of exploration and use of 'the Moon and other celestial bodies' includes the right of 'free access to all areas of celestial bodies', including Earth's orbits. [...] The term 'access' should be understood very broadly. States are barred from preventing the free access of other States or from creating difficulties in this respect.[69] However, the indiscriminate use of outer space without taking into consideration the sustainability of the missions and environmental impacts of exploring outer space might hinder the principle of freedom of exploration and use, an infringement of Article I of the OST.

Today, 50 years later, some regions of the Earth's orbital environment have reached critical object concentrations. As a consequence, several national guidelines, policies or space laws require that the launch, deployment, operation and end-of-life disposal of a space system is performed in accordance with practices that limit the growth of the space debris population. Compliance with such 'best practices' will be mandatory for the safe end-to-end conduct of future satellite constellation projects, with the aim of leaving a 'near-zero' footprint. Achieving sustainability in the uses of outer space will require a solid regulatory framework, which, as explained, does not seem to be the case either internationally or domestically. This concern was already made clear before UNCOPUOS established its agenda item on long-term sustainability. The discussion on the mitigation of space debris sparked that debate.[70]

A new and broader approach was provided with the concept of 'space traffic management' (STM) in combination with space situational awareness (SSA). Space Traffic Management (STM) is considered a 'concept to provide a framework for the safety, stability and sustainability of space activities'.[71] This concept has been

---

[67] Jahku 1992, pp. 206–207. This is also known as the 'cascade effect' or 'Kessler effect'.

[68] *Cf.* Arts. I, II, IX, Outer Space Treaty.

[69] Lachs M, Masson-Zwaan TL and Hobe S, The Law of Outer Space an Experience in Contemporary Law-Making (Martinus Nijhoff Publishers 2010) 45.

[70] https://www.un.org/sustainabledevelopment/development-agenda/.

[71] Schrogl K.-U., Jorgenson C., Robinson J., Soucek A., *Space Traffic Management - Towards a Roadmap for Implementation*, Paris, IAA, 2018; Perek L., Traffic Rules for Outer Space, International Colloquium on the Law of Outer Space (IISL), 1982, 82-IISL-09.

defined as 'the set of technical and regulatory provisions for promoting safe access into outer space, operations in outer space and return from outer space to Earth free from physical and radio-frequency interference'.[72] The term 'space traffic' was already used in the 1980s, but a more detailed discussion was initiated only in the late 1990s and culminated in a study by the International Academy of Astronautics of 2006.[73] This study became the first truly comprehensive approach to shaping a new order for the uses of outer space.

Unfortunately, it did not build on an evolution of the current framework but instead made a bold step in addressing the uses of outer space: the concept of space uses as a traffic system. While at that time many observers smiled at the expression and the concept of STM, 5 years later this approach is being seriously debated not only by academics but also by governmental (including military) and private actors in space and has even found its way into the legislation of the leading space power. STM has, therefore, to be regarded as a major approach in pursuing sustainability.[74]

6.5.5 Conclusion

When taken into consideration as a regulatory regime, all these approaches seem to address the problem of space debris rather broadly. However, the NewSpace has new legal challenges, and the sustainability of space activities is certainly one of them. It requires new rules and a new legal approach that is built together with and that answers the needs of the private sector, is needed despite the constraints of the space treaties and in compliance with them. Even though the era of space law treaty making has mostly passed, space and non-spacefaring nations need to come together once again and come to a consensus on how to best address the issue of space debris and the sustainable use of outer space. Otherwise, after 60 years of space exploration, those who have not had the chance of reaching the stars might never be able to do so in many generations to come.

Vinicius Aloia holds an LL.B Bachelor of Laws from the Catholic University of Sao Paulo (PUC-SP) and is a member of the Brazilian Bar Association. Vinicius is currently pursuing an LL.M degree in International Business Law at the University of Helsinki. In his studies, he focuses on space law, particularly the commercial uses of outer space and the legal and regulatory challenges faced by the commercial space sector. Vinicius first got involved with space law when he became a member of the European Centre for Space Law. He also took part in the 27th ECSL Summer Course on Space Law and Policy and is particularly active among the space law community in Europe and Brazil. vinicius.aloia@helsinki.fi.

---

[72] Jorgenson C., Lála P., Schrogl K.-U., *Cosmic Study on Space Traffic Management*, Paris, International Academy of Astronautics (IAA) 2006.

[73] Ibid.

[74] Brünner and Soucek (n 4) 608.

Chapter 7
Answering an Orbit Full of Questions: A Proposed Framework to Provide Legal Certainty on the Current and Future State of the Law Regulating Satellite Constellations

Anton de Waal Alberts

Abstract The First-Generation Space Law system brought certainty in relation to many types of space activities within the Cold War era. Over and above the five space treaties, several other treaties, guidelines and practices developed to establish order within the domain of satellite-related space activities. These rules and guidelines are divided into hard law and soft law categories, depending on their enforcement status. During this time, the satellite industry was dominated by governments and large private contractors and companies feeding off government projects and monopolising the little space allowed for private operations. However, the New Space era is fast changing the landscape with many new private role players having entered and many still accessing the newly enlarged space market. The speed with which this is taking place is giving rise to many new legal questions that the First-Generation Space Law system simply was not designed to deal with. Accordingly, more is expected from space law to provide certainty, which brings into sharp focus the need for the establishment of a Second-Generation Space Law system in general, but more specifically also for the satellite industry. This study endeavours to create a framework to provide certainty on the current and future state of the law that will allow for the prioritisation of issues to be dealt with as New Space changes the satellite universe.

A. de Waal Alberts (✉)
Parliament of South Africa, Parliamentary Portfolio Committee on Trade and Industry, Cape Town, South Africa

© Springer Nature Switzerland AG 2019
A. Froehlich (ed.), *Legal Aspects Around Satellite Constellations*, Studies in Space Policy 19, https://doi.org/10.1007/978-3-030-06028-2_7

7.1 Introduction

The first human-made object in space was a little USSR satellite called Sputnik 1, whose presence in space in 1957 amazed the world. Today space around planet earth is filled with an array of different types of satellites with different functions. The Cold War era gave rise to a satellite industry focused on national interest and security but also established the platform for a more interconnected civilian world. Telecommunications, communications and broadcasting services originated from and facilitated by satellites created a cloud of data and information dispersed to populations around earth, thereby giving rise to an emergent sense of a planet shared by everyone as opposed to the legacy of one nationality, people or group in conflict with another. At the end of the Cold War, it took only a few years before the Internet and other communication services provided via satellites, such as cell phones and GPS, became a normal part of daily existence.

The entry of the New Space era is currently witness to an explosion of new private role players in the satellite industry, all competing to provide legacy services more cost-effectively or new services like the operation of the Internet of Things. These events have superseded the ability of the First-Generation Space Law (FGSL) to regulate and deal with new issues arising from this increased form of space activity. There is indeed a need to understand the current state of space law in relation to satellites and the identification of new issues that deserve regulation of some sort, as well as the development of the Second-Generation Space Law (SGSL) for New Space, where needed.

This study endeavours to provide a framework to understand current and future space law in general and particularly an adapted framework for the satellite industry.

7.2 History of Satellites

7.2.1 Cold War Satellites

The Cold War satellites developed around the impetus of the space race between the 'Communist East' and the Free-market West'. This led to an explosion in research and development and gave rise to the launch of satellites generally large in size and the expansion of capabilities from the humble signals of Sputnik 1 to remote sensing for security and commercial reasons, telecommunications and broadcasting, amongst others. These satellites were initially designed, built, launched and controlled by the state in both the East and the West; however, the United States began—and later followed by other Western states—contracting private companies to develop satellites under the authority of the government.

This process ushered in the era of information, communication and telecommunications (ICT). Devezas et al. indicate that while the Cold War was waging, 'at the same time communication companies began to delve into the still unexplored field of global communications, and again synergistically three main sectors of human activities were completely transformed by the advent of rockets and related space activities: military (the menace of global destruction), entertainment (global TV broadcasting), and information exchange (IT technologies)'.[1]

Satellite development, launch and operation have constituted the bulk of space activities since Sputnik 1. Devezas et al. provides an interesting analysis of the size and cycles of the satellite industry during and at the end of the Cold War[2]:

- The total intensity of space activity reached a peak in 1968 of which satellite launch activity constituted 82%. This also indicated that other space activities took place, significantly manned space flight.
- For almost three decades between the 1960s and the 1980s, an average of 150 satellite launches per annum took place.
- After 1989 the satellite launches decreased mostly due to the implosion of the USSR whose satellite releases dropped from an average of 113 (from 1970s to 1990s) to 35 per annum.

The entry of private players changed this slump, as will be discussed *infra*.

7.2.2 New Space Satellites

Shortly after the end of the Cold War, during 1998, a strong peak of satellite launches took place due to the strong investments and growth in the production of commercial satellites from 1995. This lasted until 2002. Devezas et al. indicate that launched commercial satellites grew in this time from an average of 5 per annum to 42. This decreased from 2003 to an average of 15 satellites launched per annum.[3] The following important inferences are also made:

- In the mid-1990s, civilian satellite missions overtook the number of military satellite missions in dramatic fashion ushering in the New Space era shortly after the end of the Cold War.
- Since the mid-2000s the number of satellites launched and those in orbit started increasing again even though launches itself did not increase significantly. The reason for this trend is that increasingly more satellites were launched from the same launch vehicle.
- Launches are set to increase as new, more able, efficient and even smaller launch vehicles are developed for mass launches of small satellites, amongst others.

---

[1] Devezas T *et al.* 2012. The struggle for Space: Past and future of the space race. *Technological Forecasting & Social Change*. Elsevier Ltd.

[2] See note 1.

[3] See note 1.

The last trend has increased dramatically with the advent of small satellites. Small satellites are increasingly developed to provide commercial applications and services in competition with larger satellites even while becoming smaller still due to technological developments in other areas that provide off-the-shelf solutions and technology, like cell phone technology. Today small satellites are becoming smaller still with more applications and uses that can be accessed on the global market. Small satellites will soon be released in clusters or constellations around earth in low earth orbit, which can provide global services that are currently performed by geostationary satellites. These constellations are currently being launched for less ambitious services, but the trend is towards more robust technologies that can compete with the larger legacy satellites.[4]

7.3   Satellite Systems

Wood and Weigel classify satellites in accordance with size and use[5]:

- **Legacy commercial and government satellites**: their size is greater than 1000 kg with services relating to high-resolution earth imaging, scientific measurement, communication, positioning and timing, reconnaissance and surveillance, with a lifespan often greater than 10 years. These satellites can operate in the geostationary orbit 36,000 km from earth.
- **Small satellites:** they approximately size 100–1000 kg with services relating to earth imaging and scientific measurement in lower orbit, with a lifespan of 3–7 years.
- **Pico-, nano- and micro-satellites:** their size is less than 100 kg with services relating to technology demonstration, mission science, non-real-time communication and education in low earth orbit, with a lifespan often less than 1 year. Micro-satellites are between 10 and 100 kg, nano-satellites between 1 and 10 kg and pico-satellites less than 1 kg. A form of nano-satellite is known as a CubeSat, which conforms to a specific size of 10 cm³ with a mass of about 1 kg.

7.4   An Orbit Full of Questions: New Space Issues

The new post-Cold War multipolar world combined with international economic development in poor countries combined with more accessible space technology has resulted in a dramatic increase in international space activity. Devezas et al. illustrate the increase in the establishment of national space agencies with 30

---

[4] Wood D, Weigel A. 2014. Architecture of small satellite programs in developing countries. *Acta Astronautica*. Elsevier Ltd.
[5] See note 6.

created in the period between 1991 and 2010, which underscores the entry of more states into the space arena. The New Space activity is multiplied when private players and their activity are taken in account.

Some of the known practical issues that arise from this increased activity can be listed as follows:

- More launches are set to take place and greater constellations of large and small satellites released. This will increase both aerospace and outer space activity.
- Space traffic will increase dramatically, resulting in higher probabilities of in-space collisions and space debris.
- The current trend of the short life cycles of small satellites that may be abandoned without de-orbiting plans may also contribute dramatically to the multiplication of space debris.
- Ease of access to space technology can result in many private undocumented launches taking place without the knowledge of the state from where the launch takes place or UNCOPUOS.
- For the same reason, 'stateless' launches may take place from unregistered ships in international waters or islands not under any state jurisdiction.

These issues lead one to the following orbit full of questions:

- Are proper aerospace and outer space traffic management in place to ensure safe launches and clear orbits?
- Are proper rules in place to ensure orbits are managed and space debris mitigation measures are in place?
- Are proper rules in place to ensure liability in case of damage due to aerospace and outer space collisions?
- Are measures in place to ensure that space debris is not only mitigated but that space is also cleaned up?
- Are measures in place that ensure global control of private launches and satellite orbital releases?

These questions will have to be answered at some point. In order to establish whether the current FGSL provides answers and whether the SGSL will have to be developed, a framework of analysis is proposed below.

7.5 Frameworks for Space Activities

The design of the envisaged general and satellite-specific framework was inspired by Kish's Law of International Spaces.[6] Kish argues logically a separation of the law between national law and international law and makes a case for similar rules in public international spaces, like the polar regions, international waters and outer

---

[6] Kish J. The Law of International Spaces. Leiden: Sijthoff, 1973.

space to be applied as it relates to delimitation, the prohibition of territorial sovereignty, and jurisdiction.

Building on Kish's delimitation and on the author's own academic work, the framework proposed should be regarded as a concept and starting point from which further researched and refined versions should emerge over time.[7]

7.5.1   General Space Framework

The general space framework (GSF) is designed to be the default planning framework that sets out both the status of the FGSL (*de lege lata*) and the emergent or required SGSL as future law (*de lege ferenda*). As will be illustrated, the GSF can also be adapted and extended over time as new technologies allow access to new physical spaces and places. Smaller frameworks can be drawn from the GSF that provides focus on specific industries or activities, like the satellite industry, as will be discussed *infra*.

The GSF consists of the following elements:

- **Spatial Elements:**

  - **Earth-Orbital Zone:** it consists of the planet itself, the atmospheric aerospace and outer space with orbits up to the geostationary orbit;
  - **Ad Astra Zone:** this includes outer space in the vicinity of the earth's closest and most interesting celestial bodies, namely the Moon, various asteroids and other bodies, and the planet Mars.
  - **Outer Ad Astra Zone:** it includes the remainder of the solar system planets, including the Sun, and the remainder of the universe to be added piecemeal as it becomes technologically feasible to introduce human activity into these spaces and places.

- **Issue Identification:**

  - Various issues related to the spatial zone is identified and listed. The identification of issues is a subjective process. However, international consensus does exist in some respects and may arise in future on the issues that deserve attention and regulation. These issues are characterised as immediate, medium term and long term and correlates in general with the three spatial zones:

    Immediate issues arise from the Earth-Orbital Zone: examples would be space commercialisation in general, including commercial launchers and commercial satellites of all sizes and uses. Other issues of importance are space traffic management, space debris management, orbital slot ownership in all orbital planes and space data (Big Data) management etc.

---

[7]Alberts A de W. Corpus Juris Spatiales: A Cosmological Framework for Jurists. LL.M (International Law) Dissertation. 1998. University of Johannesburg.

Medium-term issues from the Ad Astra Zone: examples would be human and unmanned missions to the Moon, Mars and other close celestial bodies like asteroids, and jurisdiction, sovereignty and ownership on these celestial bodies. Mining and exploitation of resources are also of importance. Due to advancement in technology and political and commercial focus on newly planned missions, these identified issues may all soon take on an immediate character.

Long-term issues from the Outer Ad Astra Zone: Human and unmanned missions to other planets and celestial bodies in the solar system and beyond. Should new technology be developed that allows for faster space travel, these issues may also become medium term or perhaps even immediate.

– As briefly alluded to above in the examples, it is important to note some issues that have been characterised as medium term may become immediate as priorities change and space activity increases in that respect. For instance, the medium-term issue of human visitation and occupation of the Moon is currently being set to become an immediate issue, given NASA's current plans. Human visitation and occupation of Mars is set to take the same path due to SpaceX's plans.

• **Regulation:**

– **De Lege Lata:** identifies and presents a summary of the status of the law regulating an identified issue;
– **De Lege Ferenda:** focuses on any legal lacunae that may exist and presents possible future law.

Figure 7.1 provides an example of the concept-GSF in a two-dimensional visual form. This example will present the elements outlined *supra*. The concept-GSF must be understood against the background of the following notes:

• The issues identified as examples used below are subjective, but it can fairly be argued that they represent important international consensus issues.
• For the sake of brevity, only one issue was identified in each zone to be used in identifying the current and future law. More than one issue may be identified in practice.

This concept framework will hopefully assist space practitioners not to get bogged down in one area of specialisation but to build a holistic picture about the status of space activities and issues related thereto globally and the law and legal lacunae applicable to the various space activities.

One of the issues identified in the concept-GSF was that of satellites. The next step is now to focus on this issue for the design of a specific Satellite Space Framework (SSF).

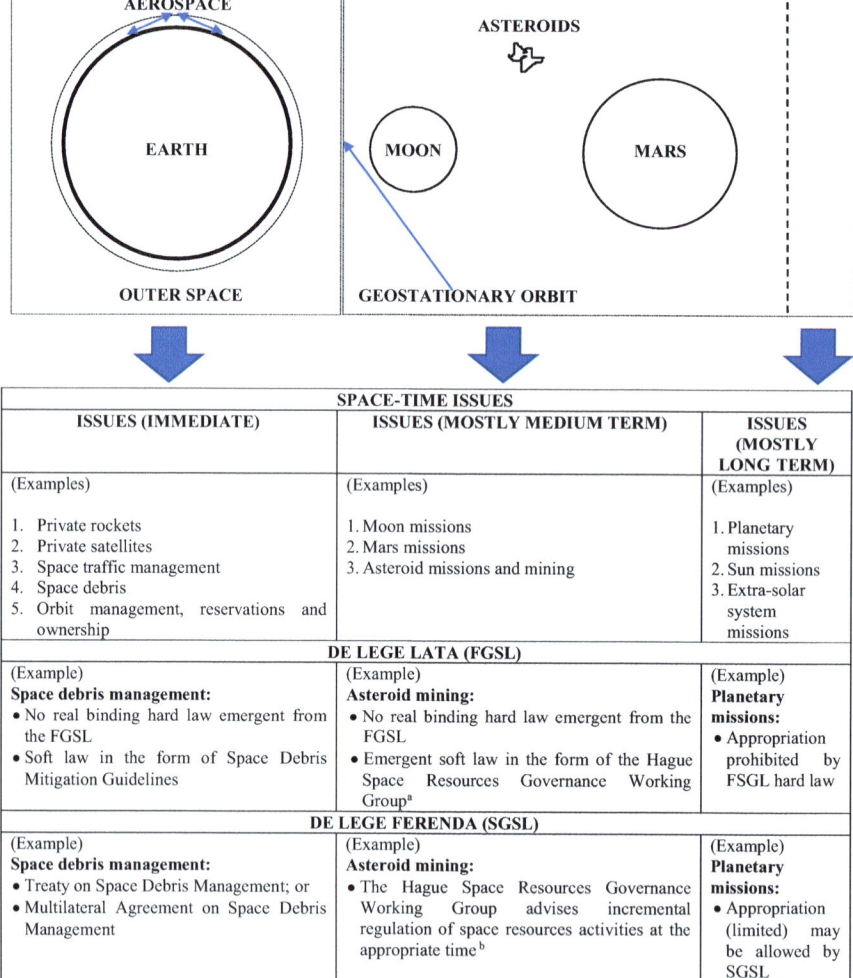

Fig. 7.1 General Space Framework (GSF)

7.5.2 Satellite Space Framework

The SSF is designed by making use of the GSF in a more focused manner. The applicable zone will for now only be the Earth-Orbital Space Zone, which extends to the geostationary orbit. As is the case with the GSF, this is subject to change that satellites may be deployed around the Moon, Mars and other celestial bodies in the Ad Astra Space Zone and even the Outer Ad Astra Space Zone.

The SSF consists of the following elements:

- **Spatial Elements:**
  - Earth-Orbital Zone: it consists of the planet itself, the atmospheric aerospace and outer space with orbits up to the geostationary orbit. In this case, with the focus on satellites, the operational area of satellite activity is identified as follows:

    Low earth orbit (LEO): satellites in this orbit are situated 500–1500 km from earth and circle the earth quickly at a rate of every 10–40 min.

    Medium earth orbit (MEO): satellites in this orbit are situated 5000–12,000 km from earth and circle the earth at a slower rate of every 2–8 h.

    Geosynchronous earth orbit (GEO): satellites in this orbit are situated 35,800 km from earth and maintain the same speed as earth's rotation, thereby establishing a locked geographical footprint on earth.

- **Issue Identification:**
  - As is the case with the GSF, identification of issues is a subjective process, but international consensus does provide more certainty. These issues can also be characterised as immediate, medium term and long term:

    Immediate issues: these issues identified are truly pressing as the intensity of satellite space activity is increasing within the New Space era. Activity has overtaken regulation, and the increasing competition between private role players will at some point lead to disputes and harm.

    Medium-term issues: Two of these issues, data and capacity sharing, can be characterised as immediate issues, as well as they have been part of the haves/have-nots debate since the start of the Cold War era. However, much has been done informally to address the issues, and this should continue in the future. At some point, however, they will again become pressing if global inequality increases. As for the issue of unregistered launches, this will probably take place on an increasing basis due to the escalation of private competition and the reluctance of business—especially start-ups—to be tied down by bureaucracy.

    Long-term issues: No long-term issues could be identified at this stage.

- **Regulation:**
  - **De Lege Lata**: in this analysis, serious lacunae have been identified that will, if not addressed, will give rise to great uncertainty, risk and disputes.
  - **De Lege Ferenda**: in this analysis, the lacunae are addressed by suggestions for regulation as part of the SGSL.

The SSF must be understood against the background of the following notes:

- The SSF is a focused version of the GSF. In this case, the whole Earth-Orbital Space Zone is relevant and used, as can be seen in Fig. 7.2 below.

EARTH-ORBITAL SPACE ZONE

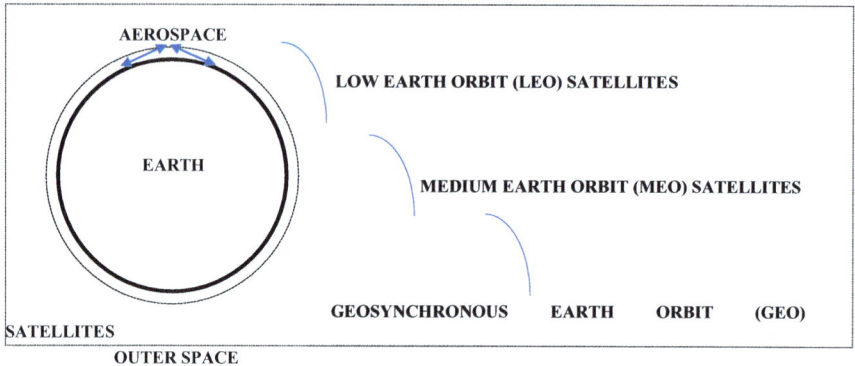

SPACE-TIME ISSUES		
ISSUES (IMMEDIATE)	ISSUES (MOSTLY MEDIUM TERM)	ISSUES (MOSTLY LONG TERM)
1. Space traffic management 2. Orbital reservation and ownership 3. Debris management 4. Liability management	1. Data sharing with poorer states 2. Capacity sharing with poorer states 3. Unregistered launches	None yet identified
DE LEGE LATA (FGSL)		
1. **Space traffic management**: no formal international regulation 2. **Orbital reservation and ownership:** no formal international regulation 3. **Debris management:** soft law by way of the Space Debris Mitigation Guidelines 4. **Liability management:** via the hard law of the Liability Convention[a]	1. **Data sharing with poorer states:** a continuing challenge with informal relationships in place for sharing of certain information 2. **Capacity sharing with poorer states:** informal and bilateral relationships between some states exist, but no global obligation exists 3. **Unregistered launches:** not a trend yet but might increase as companies defy FGSL	
DE LEGE FERENDA (SGSL)		
1. **Space traffic management:** to be addressed by soft law regulation 2. **Orbital reservation and ownership:** to be addressed by soft law regulation 3. **Debris management:** to be addressed by a Treaty on Space Debris Management or Multilateral Agreement 4. **Liability management:** refinement of the hard law where needed to bring certainty on vague matters	1. **Data sharing with poorer states:** Big Data Treaty to ensure parity in global development 2. **Capacity sharing with poorer states:** the lacuna of an international obligation can be mitigated by a treaty or a multi-lateral instrument 3. **Unregistered launches:** the FGSL places an obligation on states to register all launches with UNOOSA/COPUOS, and as such methods must be investigated to empower states to implement	

[a] UNOOSA. 2018. Convention on International Liability for Damage Caused by Space Objects. Available at http://www.unoosa.org/oosa/en/ourwork/spacelaw/treaties/introliability-convention.html (accessed on 8 October 2018)

Fig. 7.2 Satellite Space Framework (SSF)2018

- Where the focus shifts to issues closer to earth, the spatial representation will also change to represent a part of the Earth-Orbital Space Zone. For instance, if the issue is the delineation of aerospace and outer space, then the focus will be on the Earth-Orbital Space Zones without the orbital categories as set out in the Spatial Elements *supra* and as illustrated in Fig. 7.2 *infra*.

Figure 7.2 provides a two-dimensional illustration of the SSF.

The SSF set out above is a first attempt in establishing a framework to set priorities and provide a holistic understanding of the satellite industry. It should be expanded and refined over time.

Given the growing activity in the satellite industry and the emergence of complexity as a result, the SSF is designed to provide a simple reference and point of departure to understand the industry.

7.6   Conclusion: Answering the Questions

The advent of the post-Cold War era and the introduction of New Space have given rise to a historical break where private enterprise constitutes the bulk of space activity in contrast to the historical security-related space activity. Most of the private space activity consists of satellite launches and releases.

This increase in satellite activity is giving rise to a complex satellite industry where disputes will arise. It is thus important to understand the industry and know the legal framework on issues identified for purposes of planning. It is also important to know the legal lacunae that exist and to ascertain if steps are being taken to address this. It is in this regard that the proposed Satellite Space Framework can be a very useful tool.

The Satellite Space Framework is a derivative version of the proposed General Space Framework that endeavours to frame and provide a holistic overview of all space activity. The General Space Framework is designed to allow a focus on any aspect of space activities, like that of the satellite industry in this case.

Both these frameworks are first versions and are to be refined and expanded over time as is necessary. In an increasing complex global community and space industry, simple and effective points of reference in the form of these frameworks should assist space law practitioners to devise strategic plans for the complex future.

Anton Alberts is admitted as an advocate/barrister of the High Court of South Africa specialising in the legal fields of media law, ICT and space law. He is currently a Member of Parliament in South Africa and serves as a full member on the Parliamentary Portfolio Committee on Trade and Industry where he, amongst others, promotes the development of the country's space industry. He received his legal education at the University of Johannesburg, where he obtained the degrees BA (Law), LL.B and LL.M (International Law (Cum Laude)), as well as an M.Phil in Futures Studies from the University of Stellenbosch. He is a prolific researcher and has published several legal works. Anton's focus is now increasingly on space law and its development for a new era of cooperation between government and private industry.